AF307183

Springer Series in
Surface Sciences

17

Editor: Gerhard Ertl

Springer Series in **Surface Sciences**

Editors: G. Ertl and R. Gomer   Managing Editor: H. K. V. Lotsch

M. Grunze   H.J. Kreuzer (Eds.)

Adhesion and Friction

Proceedings of the Third International Workshop
on Interface Phenomena, Dalhousie University,
Halifax, N.S., Canada, August 23–27, 1988

With 58 Figures

Springer-Verlag Berlin Heidelberg New York
London Paris Tokyo Hong Kong

Professor Dr. Michael Grunze

Angewandte Physikalische Chemie, Universität Heidelberg
D-6900 Heidelberg, Fed. Rep. of Germany

Professor Dr. Hans Jürgen Kreuzer

Department of Physics, Dalhousie University, Halifax
Nova Scotia, Canada, B3H 3J5

Series Editors

Professor Dr. Gerhard Ertl

Fritz-Haber-Institut der Max-Planck-Gesellschaft, Faradayweg 4–6
D-1000 Berlin 33

Professor Robert Gomer, Ph. D.

The James Franck Institute, The University of Chicago, 5640 Ellis Avenue
Chicago, IL 60637, USA

Managing Editor: Dr. Helmut K. V. Lotsch

Springer-Verlag, Tiergartenstrasse 17
D-6900 Heidelberg, Fed. Rep. of Germany

ISBN-13: 978-3-642-74989-6   e-ISBN-13: 978-3-642-74987-2
DOI: 10.1007/978-3-642-74987-2

Preface

"Adhesion and Friction: Microscopic Concepts" was the theme of the third workshop on interface phenomena organized jointly by the surface science groups at Dalhousie University and the University of Maine. The first two workshops were dedicated to the discussion of elementary processes governing the reaction rates at surfaces and in bulk materials, i.e. adsorption, desorption and diffusion. In this third year a step towards the understanding of complicated (but practical) issues such as adhesion and friction between different materials was undertaken. The presentations and discussions focused on elementary chemical and physical processes at surfaces and interfaces relevant to adhesion, lubrication and friction and gave an account of the application of surface science methods and techniques to relevant model systems. Clearly, at the time of the conference and the publication of the proceedings the understanding of the chemical and physical mechanisms determining the interaction between two solids is still rudimentary, but the issues involved are attracting the attention of more and more scientists and are now regularly represented at scientific meetings.

The conference was held at Dalhousie University in Halifax, Nova Scotia, Canada. The facilities provided an ideal setting for the meeting and lively discussions. On behalf of the participants, we would like to express our gratitude to the staff at Dalhousie University for making our stay so pleasant and memorable.

The workshop was supported by the Air Force Office of Scientific Research, the Office of Naval Research, the Surface Science Division of the Canadian Association of Physicists, the University of Maine and the Provincial Government of Nova Scotia, Canada. The views, opinions and/or findings contained in this report are those of the authors and should not be construed as an official position, policy or decision by any of the above agencies, unless so designated by other documentation.

We would also like to thank the following companies for their donations to help sponsor graduate student attendance:
Alcan, Canada
Howell Laboratories
McAllister Technical Services
GTE Laboratories
B.P. America, Inc.

Heidelberg, Halifax
April 1989

M. Grunze
H.J. Kreuzer

Contents

Friction at Weakly Interacting Interfaces

G.M. McClelland

IBM Research Division, Almaden Research Center, San Jose, CA 95120, USA

1 Abstract

For two atomically flat surfaces which interact only by van der Waals bonds, sliding should occur without wear. Possible mechanisms of energy dissipation during slow sliding are described for such a system. According to both an independent oscillator model and the Frenkel-Kontorova model, the static friction should be zero when the solids are incommensurate. A two dimensional numerical simulation supports this conclusion. As the interfacial interaction is increased, there is a well-defined threshold for the onset of friction. Depending on the shape of the interaction potentials, mobile molecules at the interface may present an additional energy dissipation mechanism.

2 Introduction

Although the processes of friction, wear, and lubrication have been the subject of much investigation, there is little understanding at the atomic level of what occurs at a sliding interface. Traditional measurements of friction and wear coefficients can not be easily compared to an atomic scale theory, because such measurements are averaged over large heterogeneous samples.

By shrinking the contact down to a single asperity, a measurement can be better controlled [1]. The force during contact can be measured, and damage to the sample can be assessed by electron microscopy after sliding. Field ion microscopy has also been used to measure the damage to a sharp tip in contact with the sample [2]. The recently developed atomic force microscope (AFM) [3] promises to enable frictional measurements at contact areas down to a single atom. Although this goal has not yet been realized, we have recently used an AFM to detect the first atomic scale features in a frictional force. When a tungsten tip slides across a graphite [4] or mica [5] surface at loads $\simeq 10^{-5}$ N, the frictional force varies with the spatial period of the underlying atomic lattice.

Although the surface force apparatus (SFA) uses large (30 μm) contacts, the contact geometry is defined very well by using atomically flat mica or coated mica cylinders with radii of $\simeq 1$ cm. Mica sheets are almost unique in that they can be cleaved step-free over regions several mm across. Using the SFA, the shearing properties of precisely controlled layers of solids and liquids (down to a single monolayer) have been studied [6,7,8,910].

Springer Series in Surface Sciences, Vol. 17
Adhesion and Friction   Editors: M. Grunze and H.J. Kreuzer
© Springer-Verlag Berlin, Heidelberg 1989

Recent three dimensional molecular dynamics simulations involving many atoms and using realistic potentials have begun to provide a detailed picture of what happens at sliding interfaces [11, 12], including the tip-surface interface in an AFM [11,13]. In an effort to focus on key atomic aspects of the mechanism of energy dissipation at sliding surfaces, greatly simplified pictures of the dynamics are discussed in the present paper. Detailed molecular dynamics calculations and experiments on well-defined systems will determine to what extent the simple models presented here are valid.

The importance of determining how energy is dissipated during sliding friction has long been recognized. In most cases friction occurs through an adhesive mechanism, in which the frictional energy is dissipated in plastically deforming material asperities at the interface. During this type of friction, as in ploughing friction, the energy is dissipated just as it is in macroscopic deformation of a material, by the motion of dislocations during shear. However, there seem to be some cases where sliding (with friction) does occur without wear [1,14] on a single asperity. This is particularly clear for the AFM friction studies, where the same atomic scale features persist after 100 passes of the tip over the surface [4, 5]. The fact that the AFM can image surfaces with atomic resolution through the perpendicular repulsive force between a tip and a surface [15] is also convincing evidence that sliding can occur without wear.

In this paper, we discuss simple models of the friction of a sliding interface specifically chosen to involve no wear. The system involves two atomically flat solids sliding infinitesimally slowly across each other. We assume that the forces holding each solid together (covalent, metallic, or ionic bonds) are much stronger than the physical interactions (van der Waals forces, dipolar forces, hydrogen bonding, repulsive forces) by which the solids interact. We will emphasize that as long as the two surfaces are incommensurate, there is essentially no friction: the two surfaces can slide across each other without dissipating energy. This conclusion is well known in the study of many physically analogous systems involving two interacting periodicities [16], including charge density waves [17], ionic conduction [18], epitaxial growth [19,20], and adsorbed monolayers. Theoretical descriptions are usually based on the well-known Frenkel-Kontorova (FK) model [16,21,22,23.]. We emphasize this possibility of frictionless motion here, because it seems not to be generally recognized among workers in tribology, despite a recent paper by SOKOLOFF [24] which has discussed this issue.

To clarify the mechanism of energy dissipation during sliding without wear, we introduce an extremely simplified model of a sliding interface which is similar to a model presented by TOMLINSON 60 years ago [25]. This independent oscillator (IO) model is complementary to the FK model, in that it makes the opposite set of approximations in simplifying the interface. Interestingly, the IO model leads to the same qualitative conclusion as the FK model: that for interfacial forces below a certain threshold size, frictionless sliding occurs for incommensurate interfaces. The results of molecular dynamics calculations are then presented, which confirm the conclusions of the two simpler models.

Adsorbed atoms or molecules will nearly always be present at a sliding interface unless a special effort is made to exclude them. Such layers have been the subject of recent study in the SFA [6,7,8,910]. and undoubtedly play a role in the AFM experiments [4,5]. A simple model is presented here shows that mobile adsorbed molecules can at times act not like a lubricant but provide an additional channel for energy dissipation.

3 The Independent Oscillator Model

Figure 1a displays the system of interest, containing a lower solid labeled A and an upper solid B, reduced to the two dimensions of the plane of the figure. The lattice constants of the two solids differ, and for each solid, the layer of atoms furthest from the interface is attached to a rigid (black) support. Sliding the two solids over each other is performed by moving the A support to the right while keeping the distance between the supports constant. Experimentally it is normally the load which is specified, but the major effect of load for rough surfaces is to vary the contact area, an irrelevant effect for the present system. We assume that the distance between the supports has been set initially to that of mechanical equilibrium, where there interfacial potential is minimized, and there is no net normal force between the supports. (Again, this is not to be confused with zero normal load for the usual case of rough surfaces. In our example of smooth surfaces, the interfacial interaction can be substantial, even though there is no external normal load.)

Suppose that the atoms are strongly bound within each solid by covalent, ionic, or metallic bonds, but that the interaction across the interface is weak, composed only of van der Waals, polar, hydrogen bonding, and/or repulsive forces. The intersolid bond energy will then be less than 1/10th that of the intrasolid bonds, and since these bonds also vary more slowly with distance, they break with $<<$ 1/50 of the force of the intrasolid bonds. Thus as solid A is translated along x, shear occurs at the interface, with no wear, and no rupture of the intrasolid bonds.

The static friction at the A-B interface is the force required to translate solid A across B at an infinitesimal velocity, while maintaining the boundary conditions as described above. We assume that at the boundaries the solids are attached to heat sinks at zero temperature, so that any energy generated will be dissipated.

For static friction to be present, some mechanism must convert the infinitesimally slow motion of the A support to rapid atomic vibration (heat). This mechanism must also lead to a dependence of the sign of the frictional force on the direction of motion. Such a mechanism can be identified through the independent oscillator (IO) model, diagrammed in Fig. 1b. Here solid A has been simplified to a single row of rigidly connected atoms. The row of B atoms at the interface do not interact with each other, but they are subject to an interfacial potential from atoms of A and are connected by a single flexible bond to a rigid support representing the remainder of the solid B. To model dissipation of any excitation of the interfacial B atoms, we can allow these bonds to lose energy into the B support. Despite the simplicity of the IO model, it contains all the major elements of the wearless frictional problem:

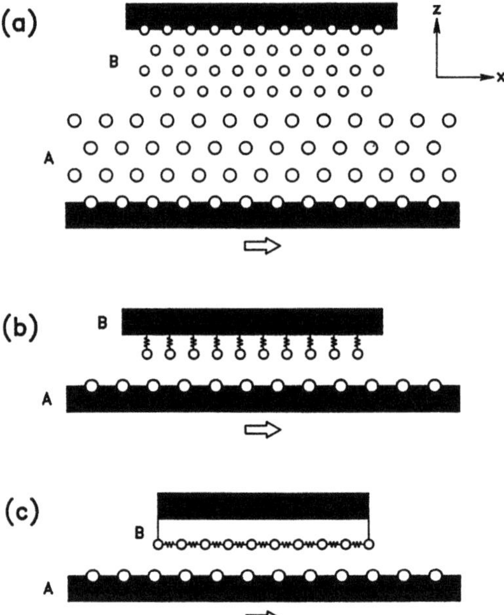

Figure 1. (a) Solid A sliding across solid B in two dimensions. (b) Independent oscillator model. (c) Frenkel Kontorova model. The black areas represent rigid supports of the atoms.

a periodic corrugated interfacial structure, movable surface atoms, and a mechanism to dissipate vibrational energy created at the surface. Our analysis is similar to that of TOMLINSON [25], who recognized long ago the importance of the problem of mechanical adiabaticity in understanding friction.

We now analyze the dynamics of the IO model in terms of the potential energy curves V_{AB} and V_{BB} governing the lateral motion of a particular B atom B_0 as A slides by it (Fig. 2). V_{BB} is a simple parabolic potential well with a curvature k_{BB}, representing the strong bond connecting B_0 to solid B. For the purpose of argument, the initially assumed form for V_{AB} shows periodic sharp peaks for each A atom; this form represents a sharp repulsive potential. (Later we introduce a weaker form for V_{AB}, which is more appropriate for the kind of systems in which we are actually interested.) As diagrammed on the left side of Fig. 2a, the net potential governing the motion of B_0 is the sum $V_S = V_{AB} + V_{BB}$; the changing shape of this sum as A is advanced along x governs the motion of B_0. These potentials are presented as effective potential functions of a single coordinate along the path of B_0, which lies mostly in the x direction. The sum V_S is shown on the left side of Fig. 2 for sequential positions of solid A as it is moved to the right; at each position of A, B_0 rests at a local minimum in the potential. If A is slid infinitesimally slowly so that V_S is changed infinitesimally slowly, the principal of adiabatic invariance requires that the classical action for the B_0 motion is conserved [26], i.e. B_0 remains in the (changing) minimum of V_S, without becoming vibrationally excited. The exception

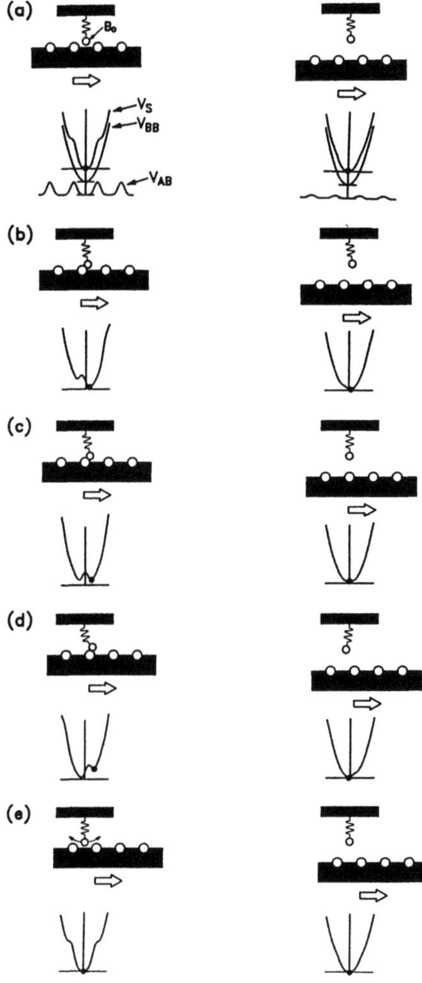

Figure 2. Motion of an atom B_0 of B in the independent oscillator model. Left and right columns diagram strong and weak interfacial interactions, respectively. The top panel diagrams the relevant potentials, while subsequent panels illustrate the response of B_0, represented by a black dot on V_S, to progressive sliding of A, as determined by the potential sum V_S plotted below each diagram.

occurs at a point between Fig. 2d and Fig. 2e where a local minimum disappears. Then B_0 must fall abruptly to the potential bottom, becoming vibrationally excited. This vibrational energy is then dissipated irreversibly into the solids. Effectively, the potential V_{AB} has "plucked" the harmonic V_{BB} bond.

Importantly, the directionality of the frictional force is a consequence of the double minimum in V_S induced by V_{AB}. For example, for the A position of Fig. 2c, if A has been moving to the right as just described, atom B_0 is pushed to the right, thereby exerting a force on solid B to the right. If A has been moving to the left, B_0 finds itself in the left minimum, and the resulting forces are reversed.

Although our interest is in systems with interfacial interactions much weaker than intrasolid interactions, this is clearly not the case for the potentials just used, for to induce a local maximum in V_S, the downward curvature at the peak of V_{AB} must have been greater than the force constant k_{BB}. An example where V_{AB} is very weak compared to V_{BB} is illustrated in the right column of Fig. 2, which is identical to the left column except for the reduction of V_{AB} by a factor of 5. Here the atoms at the interface have been drawn as being further apart to model the weaker interaction of a van der Waals bond. For the right column, the upward curvature k_{BB} of V_{BB} is greater than the magnitude everywhere of any downward curvature of V_{AB}:

$$k_{BB} > -\frac{\partial^2 V_{AB}}{\partial^2 x_{B_0}} \tag{1}$$

Thus V_S is everywhere concave upward and has no maxima, and therefore has only a single minimum for any position of the solid A. No minima disappear as A slides, there is no means for atom B_0 to become excited, and no energy can be dissipated at the interface. In other words, there is no friction. As far as B_0 is concerned, sliding A adds a slowly changing perturbation to the harmonic oscillator potential V_{BB}. In the limit of infinitesimally slow change, B_0 remains vibrationless, even though the equilibrium position changes.

If there is no energy dissipation, the net lateral interfacial force should be conservative, not depending on the direction of motion. Unlike the case for strong interaction, for weak A - B interactions, the displacement of B_0 is the same for either direction of A motion, and simply oscillates about zero as A is translated. Because there are no longer two minima positions for B_0 there can be no directional dependence to the force.

From another point of view, it is clear that in sliding two solids across each other, breaking interfacial bonds (even van der Waals bonds) and distorting intrasolid bonds at the interface requires energy. For the weak interaction case satisfying (1), energy is returned through the reversible formation of new bonds and the relaxation of the distortion. For the strong interaction case, the energy is not returned; the new bonds are formed and the relaxation occurs instantaneously, so that the energy is not regained as mechanical energy but dissipated to the rest of the solid as phonons.

To further emphasize the effect of changing potential shapes on the frictional force, suppose that the interfacial potential has the form

$$V_{AB} = C_{AB}\left(\frac{a_A}{2\pi}\right)^2 \cos\left[\frac{2\pi}{a_A}(x_{B_0} - \delta_A)\right], \tag{2}$$

which has a period a_A and a maximum downward curvature of magnitude C_{AB}. Then whenever C_{AB} is less than k_{BB}, there is no maximum in V_S and no frictional energy is dissipated. As C_{AB} increases from zero, friction remains zero until it reaches k_{BB}, from where the friction begins to increase with C_{AB}

Now consider the frictional force added over many B atoms, as indicated in Fig. 1b. To compute the lateral force on B, we sum over all the atoms of B. Using again the potential V_{AB} of (2), we suppose that there are N atoms in B, with a natural spacing of a_B. When $C_{AB} << k_{BB}$, the B atoms have approximately this natural spacing. The total lateral force on solid B is then

$$F_{Tx} = \sum_{n=0}^{N-1} C_{AB} \cos\left[\frac{2\pi}{a_A} (x_{B_0} + na_B) \right]$$

$$= \frac{C_{AB}}{2} Re\left[\exp\left(\frac{i2\pi x_{B_0}}{a_A} \right) \frac{\exp\left(\frac{2\pi i N a_B}{a_A} \right) - 1}{\exp\left(\frac{2\pi i a_B}{a_A} \right) - 1} \right] \tag{3}$$

where x_{B_0} is the position of the end atom of B. As long as a_B/a_A is not an integer, this sum is on the order of C_{AB}. Thus regardless of the length of the interface, the total force is on the order of C_{AB}, the force required to move a single atom across the interface. For typical systems, this is less than 10^{-8} grams, so practically speaking, the static force is zero. The forces on the individual atoms are adding out of phase, so that there is no buildup when summing over all the B atoms. REISS has shown that the same phenomena holds when summing over a two dimensional interface between two three dimensional solids [27].

Something very different happens when a_B/a_A is an integer (and also, if the motion of the B atoms and a non-sinusoidal interfacial potential is considered, whenever a_A and a_B are in the ratio of simple rational numbers). In this case, the interfacial forces add in phase along the interface and can be very large. Consider an interface N atoms long where $a_A = a_B$; then the lateral force is N times the force for one atom, so a sizable force is required to move A. Now in our simple model of Figure 1b, the B atoms simply move independently, uninfluenced by the coherence of the interfacial force. In a real elastic solid (Fig. 1a), in which the interfacial atoms are not attached to a rigid support but interact with one another, a force F_x placed on each of N neighboring atoms will generate a lateral distortion much greater than that of a single atom on which F_x is exerted, because the force applied to many atoms need generate only small strains, not locally large strains. For this large distortion, involving the coherent displacement of many atoms, the effective k_{BB} will become smaller than the effective C_{AB}, so that frictionless motion is no longer possible. A discontinuous motion, dissipating energy, will occur, which in a real solid will take the form of hopping of interfacial dislocations. This qualitative argument indicates the importance of including correlated motion of neighboring interfacial atoms, an effect discussed in the next section.

4 The Frenkel-Kontorova Model

Although many simplifications were made in the previous section, the most serious was the almost complete neglect of the structure of the solids. The interfacial B atoms interact not simply with a rigid support as in the IO model; they interact with other B atoms, which are themselves acted on by A atoms. Such interactions are included in a model first proposed by FRENKEL and KONTOROVA [21], and soon developed by FRANK and VAN DER MEWE [22] and later by many others [16]. SOKOLOFF has also recently discussed the application of this model to mechanical friction [24].

The Frenkel-Kontorova (FK) model, diagrammed in Fig. 1c, describes the mechanics of a row of B atoms constrained to a line and allowed to interact with each other through a harmonic potential between each atom and its neighbors,

$$V_{BB} = \frac{k_{BB}}{4} \sum_n (x_{Bn} - x_{Bn-1} - a_B)^2. \tag{4}$$

The average spacing between the B atoms is required to be a_B, but the B atoms are not anchored to some rigid substrate as in Fig. 1b, but interact only with their nearest neighbors. Although other boundary conditions can be applied, for our application, the average spacing a_B is ensured by anchoring the ends of the B chain to the rigid support. To facilitate comparison to the IO model already discussed, the normalization of V_{BB} has been chosen so that if all atoms but one are held fixed at their equilibrium position, the force constant for moving the atom is $\partial^2 V_{BB}/\partial^2 x_{Bn} = k_{BB}$. Each of the B atoms interacts with A (again assumed to be rigid) with the sinusoidal potential of (2). Note that while the IO model does not promote the coherent motion of neighboring B atoms, the FK model allows coherent motion to happen more easily than it would in a real solid.

In the FK model, there is a competition between the B - B interaction, which acts to keep the B atoms spaced by a_B, and the A - B interaction, which acts to space the B atoms at the minima of the A - B potential, which are spaced by a_A. When the parameter C_{AB} is zero, the B atoms are evenly spaced at a_B. For most cases where $0 < C_{AB} << k_{BB}$, the B atoms are slightly displaced from equal spacings of a_B, the displacement depending on the local phase of A and B. For $C_{AB} >> k_{BB}$, however, a qualitatively different behavior occurs. For this case, the B atoms are forced to be locally in phase with the A lattice, lying in the minima of the potential (2). but between such regions there are "misfit dislocations" which have the effect of allowing the average spacing of B to be a_B. The lattices are said to be "pinned" together.

We are interested in the force per B atom required to move the B lattice with respect to the A lattice. In the strong interaction case ($C_{AB} > k_{BB}$), this force per atom is substantial, but the force per atom may be considerably less (as in dislocation motion in the interior of perfect crystals) than the force required to move a single B atom in the A potential. However for smaller values of C_{AB}, the force required per atom to move the B layer is zero for an infinite system [16,23]. In other words, there is no friction (at zero velocity) between A and B. For a finite system, the force required to translate all of A is conservative (independent of the direction of motion), oscillating about 0 as A is translated, with a magnitude on the order of that of the force required to translate a single atom of A along B. For this weak interaction case, we have a situation very much analogous to that discussed in Section 3, the A spacing resulting in an out of phase sum over the B interaction. For the strong interaction case, the A lattice pulls B locally into phase with it, and to move A across B, the B atoms must "jump" between adjacent minima of A.

At what critical relative value $(C_{AB}/k_{BB})_c$ of the interaction and spring potentials does the transition from free sliding to frictional behavior occur? The value depends strongly on a_B/a_A. If a_B/a_A is near a simple rational number, $(C_{AB}/k_{BB})_c$ is quite small, because V_{AB} acts in phase over many atoms of the chain. The highest values of $(C_{AB}/k_{BB})_c$ range up to $0.2\pi^2/4 \simeq 0.5$ [23] and are for a_B/a_A values far from a simple rational number. The highest values of $(C_{AB}/k_{BB})_c$ correspond to the least amount of correlated motion among B atoms, and is expected to agree most closely with the value from the simple IO model, which is unity. That a smaller $(C_{AB}/k_{BB})_c$ is obtained for the FK than for the IO model is in accord with the fact that, if B atoms neighboring a particular B atom are allowed to relax in response to its motion, its effective force constant will be decreased. In view of the large differences between the two models, the agreement of the predicted $(C_{AB}/k_{BB})_c$ values to within a factor of two is satisfying.

Although further quantitative comparisons of the FK and IO model are probably unwarranted, it is remarkable that these two models give the same qualitative results for understanding friction. The FK model describes a much richer physics, because it includes the ability of the V_{AB} to locally distort the B solid into registry with it. As a model of the interface between two semi-infinite solids, the FK and IO models seem equally inadequate, but in different ways. The FK model neglects the interaction of the interfacial layer of B with the remainder of B. The IO model includes this interaction in a simplified way, but neglects the the interaction of neighboring interfacial atoms of B.

5 Model Calculations

The fact that the simplistic IO and FK models both predict frictionless sliding, even though they make complementary approximations in modelling the interface, strongly suggests that a realistic model would give the same result. We have performed a small set of numerical computations on arrays of atoms similar to that of Fig. 1a, and do indeed find frictionless sliding.

Close packed lattices are used, with additive central Morse potentials. The bond energies for each lattice are identical, but their lattice parameters satisfy $a_B/a_A = .8451$. Interfacial bond energies between A and B atoms is 1/10 of the intrasolid bond energies, and the A-B Morse bond length is 1.5 times the average of the intrasolid bond length.

Except for the numbers of atoms, the calculations simulate the geometry of Fig. 1a. The bottom solid is 75 atoms long and 6 atoms high, while the top solid is 60 atoms long and either 6 or 18 atoms high. To begin the calculations, the distance between the two parallel supports is adjusted (while relaxing the atoms) until the force normal to the interface is zero. Rather than compute the atomic dynamics by solving Newton's equations, an approximation to adiabatic advancement of the bottom support was performed. For each cycle of the calculation, the support was advanced by $\Delta x_A = a_A/20$. Then all atoms of A and B except those imbedded in the supports are slowly relaxed along their force gradients. After each $a_A/20$ advancement of the support, several 100 iterations are required to find a mechanical equilibrium for the atoms. After finding this equilibrium, the *total* shear force along the interface is computed, and x_A is again advanced.

For the parameter set described above, we find very low shear forces. Furthermore, the displacements of all atoms (after relaxation is performed for each x_A step) are smooth functions of x_A. And both the displacements and the shear force are reversible with reversal of the direction of sliding. Under such conditions, the forces and atomic positions calculated with this method are those which would be computed by solving Newton's equations beginning at zero absolute temperature and advancing the A support infinitesimally slowly.

Figure 3 presents the total shear force (the total lateral force exerted on each support) as a function of the displacement of the A support. The displacement x_A is normalized to the A lattice constant, while the total shear force F_{Tx} is normalized to the force F_{AB} required to break a *single* interfacial Morse bond. In the figure, $x_A = 0$ represents the point at which the vertical center line of the two solids coincide. If the lateral force is to be reversible, it must by symmetry be zero at this point. The fact that it is slightly different from zero is a measure of the incompleteness of the relaxation procedure. The plot only extends over half of the A lattice constant, but by symmetry it must have a period of a_A and be inversion symmetric around the point (0.5,0). (The period is a_A and not a_B, because, from the point of view of the sliding interface, A is infinite (longer than B), and symmetric to translation by a_A, while B, having ends, has no translational symmetry.)

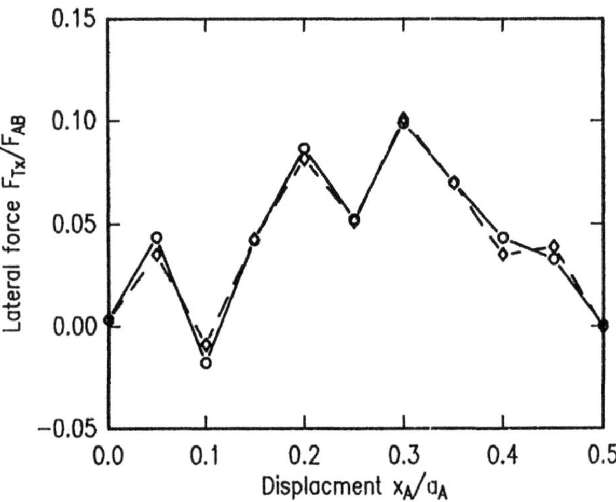

Figure 3. Lateral force (normalized by interfacial bond strength) *vs.* displacement (normalized by a lattice constant) of a sliding interface. The values are from a numerical simulation of a geometry similar to that of Fig. 1a. The circles and triangles result for top solids 18 and 6 atoms thick, respectively.

The remarkable feature of this plot is that the lateral force never exceeds $F_{AB}/10$. In other words, unlimited sliding along this 60-atom-long interface requires a *total* lateral force only 1/10 of that required to break a *single* interfacial Morse bond. This is in fact about the force required to push an adsorbed atom of A along the surface of B. In agreement with the discussion of Section 2, the magnitude of this force does not generally increase as the interface is made longer. As seen from Figure 3, the lateral force is changed only slightly by changing the height (i.e., thickness) of the upper solid from 18 to 6 layers. This suggests that the results would be the same for an extended two dimensional solid.

For flat surfaces, we expect that even a very large external compressive load across the interface will have little effect on the friction, because the interfacial forces are already quite high compared to any externally applied force. By moving the supports closer together, a very large external load sufficient to compress the solids by 1 % was applied. Sliding remained reversible and non-dissipative, and the lateral forces increased only 20 % above the force for no load.

6 Effect of Mobile Molecules at the Interface

We now consider qualitatively the effect of molecules trapped at the interface between two flat, weakly interacting surfaces. Unlike the case of boundary lubrication, in which molecules are strongly bound to one or the other of the solids, we assume the molecules interact weakly with both surfaces and are not permanently bound to particular sites. How can such molecules contribute to energy dissipation (friction) at the interface?

The model is presented in Fig. 4. Although mobile molecules are often present in continuous layers, we make the drastic approximation of neglecting interactions between the molecules. It is assumed that each solid is much stiffer than the inter-solid bonds and the molecule-solid bonds. The solids undergo no deformation, and maintain a constant relative spacing. Assuming an infinitesimally low temperature, the motion of an interfacial molecule is taken as effectively one dimensional along the interface, as it follows the path of lowest energy. For a given relative position of the surfaces, the potential energy for motion of the interfacial molecule M is assumed to be the sum of a potential to each solid $V_S(x_M) = V_{MA}(x_M) + V_{MB}(x_M)$. As solid A is displaced, the M potential can be written

$$V_S(x_M) = V_{MA}(x_M - x_A) + V_{MB}(x_M), \tag{5}$$

where x_A is the displacement of solid A. To determine the molecular motion as A slides, we apply in Fig. 4 the scheme used in Fig. 2, following the variation of V_S with x_A . In the left column of Fig. 4 we have chosen the V_{MA} and V_{MB} to have sharply curved maxima separated by broad minima. Beginning at $x_A = 0$ the adsorbate molecule is in the minimum of V_S. Assuming the temperature is so low that thermal activation is impossible, as A slides from $x_A = 0$ to $x_A = 3a_A/4$, the energy of this minimum increases, carrying the adsorbate molecule with it. Before $x_A = 7a_A/8$ is reached, however, the local minimum disappears and the adsorbate slides abruptly down to a lower minimum of the potential. This is a sudden motion,

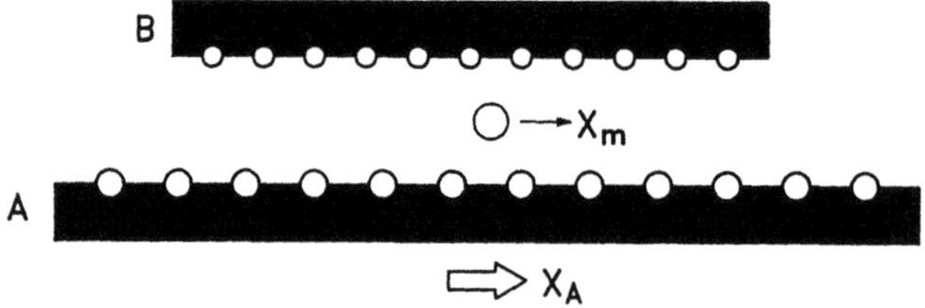

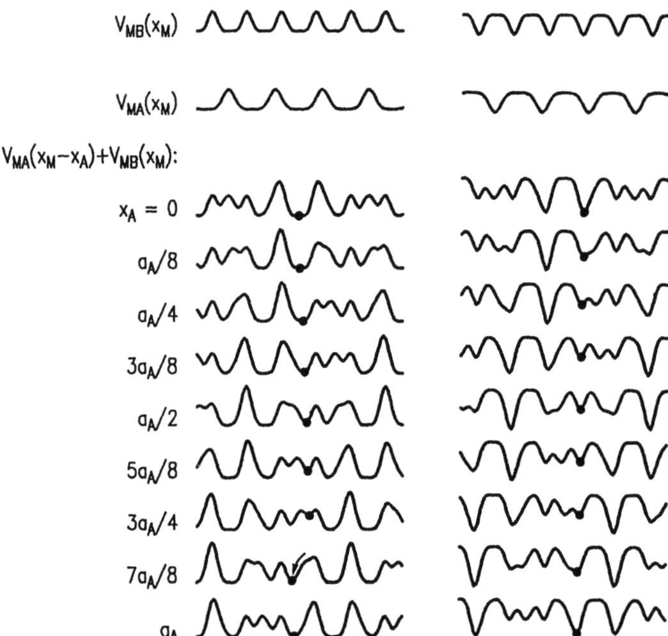

Fig. 4. Motion of mobile molecule at a sliding interface. As the lower solid translates, the plots indicate the change caused in the total potential for molecular motion. For the left column the potential minima from each solid are more weakly curved than the maxima, while for the right column, the opposite holds.

and the energy change ΔE of the adsorbate molecule is released into phonons, (heat) at the interface. The molecule has become laterally "trapped" between the repulsive parts of the potential, but subsequently "pops out", releasing energy. As we complete the cycle, so that the total displacement of A is one lattice constant, we regain a small amount of molecular potential energy, which exerts a force along the displacement direction (partially cancelling forces opposing the sliding exerted elsewhere along the interface), but the overall loss is still ΔE.

As for the IO model, energy dissipation is not a general property of the model we have constructed, but depends on the exact shape of the potential. From an analogy to Fig. 2, it can be seen that the dissipation occurs because the maxima of V_{MB} and V_{MA} are sharper (stronger curvature) than their minima. The opposite case (sharper minima than maxima) is illustrated in in the right column of Fig. 4, where in fact the potentials are simply the negatives of those of the left column. In this example we see that there is no sudden release of the interfacial adsorbate as the surfaces are slid because local minima which evolve from the deep minima at $x_A = 0$ never disappear as x_A varies. The dissipation of energy requires that a metastable minimum in V_S disappear as the surfaces are slid. Potentials such as those in the left column of Fig. 4, have this property, while potentials such as those in the right column do not. (There are some high shallow disappearing minima in the right hand column, but these are never populated during sliding by molecules which start in the lowest minima)

In conclusion, a weakly bound adsorbate can in principle *increase* the frictional force between two solids, even when the interaction of the adsorbate with either solid is weaker than the interactions of the solids themselves. Whether or not energy is dissipated by a weakly bound adsorbate depends on the shape of the potential binding the adsorbate to the substrate, not on the magnitude of the potential.

The most glaring inadequacy in the present discussion is that we have treated independent interfacial molecules. In most cases, these molecules are expected to form layers with their own characteristic spacing. There is a potential for very rich behavior dependent on the ordering of the molecular layer with respect to both solids.

7 Summary and Discussion

We have considered the problem of slow sliding without wear at the atomically flat interface between two solids. If the intersolid interactions at the interface are substantially weaker than the intrasolid forces and the solids are incommensurate, the motion of the atoms at the interface is slow and reversible, their is no dissipation of energy, and the static friction is zero. For commensurate or more strongly interacting solids, the atomic motion at the interface becomes sudden and irreversible, energy is released into phonons, and the frictional force is non-zero. This same picture emerges from three models of the problem: the independent oscillator model, the Frenkel-Kontorowa model, and a numerical model of a two-dimensional lattice. A similar picture holds for energy dissipation by mobile molecules at the interface, except that for this problem (at zero temperature) the shape of the molecule-surface potential is more important than its depth.

Friction is a process in which motion along a slow external coordinate is converted to heat, which is rapid vibration at the atomic level. Referring back to Fig. 1a, motion of the A support can be regarded as variation of a parameter of the Hamiltonian governing the motion of the atoms of the A and B solids. According to the principle of adiabatic invariance, an infinitesimally slow (adiabatic) change in such a parameter will in many cases not excite any of the action variables of the system [26]. The cases of zero friction we have discussed are examples of this principle. However, the principle is inapplicable when an infinitesimal change in the external parameter creates a *qualitative* change in the Hamiltonian. This is just what happens in the "strong" interfacial interaction cases described here: an incremental change in the effective potential for an interfacial atom or molecule results in the disappearance of a stable equilibrium point of the atom or molecule, with a resultant dissipation of energy.

Although the arguments presented here specifically pertained to the limit of zero sliding velocity (static friction), a system which behaves adiabatically at low velocities may show a finite friction at higher velocities due to excitation of phonons by the oscillating interfacial potential. In the absence of thermally activated processes, it is natural to compare the sliding velocity to the acoustic velocity, which is the rate at which strain relaxes in a solid. By this standard, it is clear that many cases of practical interest are in fact near the zero speed limit. This is particularly true for AFM and SFA experiments, where the sliding speed is on the order of 10^{-7} to 10^{-2} cm/s.

The prospect of creating frictionless solids is an attractive idea, but there are practical difficulties, because real solids contain defects. If such defects create a locally strong interaction across the interface, the local potentials governing the motion will behave as in the left, not the right, column of Fig. 2, and friction will result, as has been modelled by SOKOLOFF [24]. For a small defect density, the frictional force might be very small, but defects might initiate wear. Once wear has started it would likely increase rapidly, destroying the flat sliding geometry necessary for low friction. What is needed are "self-healing" tribological systems, which maintain smooth interfaces in the presence of wear.

Frictionless sliding may in fact have already been observed for large metal clusters on inert substrates. Quite large clusters are known to undergo thermally activated diffusion with anomalously low apparent energy barriers [28], an effect which has been attributed to lattice incommensurability [27].

It is tempting to equate the work done by the frictional force done to the energy of the bonds (even if only van der Waals bonds) which must be broken as sliding occurs [8,29]. New bonds are formed simultaneously, however, and the net frictional dissipation depends on what fraction of the energy released in this formation goes into promoting sliding and what fraction goes into creating atomic vibrations at the interface. In the frictionless systems discussed here, all of this released energy goes into the promotion of sliding.

The role of thermal motion has been entirely neglected in this paper, but it will have a qualitative effect when the temperature is high enough to thermally equilibrate the population of the disappearing metastable potential minima required to dissipate energy (Figs. 2 and 4). Clearly if the position of the interfacial atoms is determined by thermal motion and not by the direction of sliding, there can be no directional dependence to the lateral force and hence no friction at zero sliding velocities. In such a situation, a frictional force could develop at non-zero velocities (phenomenologically like a viscosity) when the sorting of interfacial atoms into different regions of the potential by sliding competes with thermal excitation.

Considering the important role of incommensurability in achieving frictionless motion, it is worth pointing out a difference between the interface between two identical planar solids interacting along a line (as in our models) and the interaction of two identical three dimensional solids across a plane. The three dimensional solids are effectively only commensurate when they are angularly aligned. Thus by controlling the relative orientation across the interface, the effect of commensurability can be explored. A sharply peaked angular effect has been observed in the adhesive force between two mica plates [9], and it is likely that an even stronger effect should be present in frictional forces.

Although much of what has been discussed in this paper has been stated by others before, it is hoped that by focussing attention on simple atomic models of a sliding interface, this paper will encourage more thorough theoretical and experimental studies of the atomic dynamics of friction.

I thank F. F. Abraham, D. DiVincenzo, M. Miller, E. Rabinowicz and C. Wöll for helpful discussions and S. Cohen for reading a portion of the manuscript.

References

1. For example, see N. Gane and F. P. Bowden: J. Appl. Phys 39, 1432 (1968); J. Skinner and N. Gane: J. Phys. D 5, 2087 (1972); A. Kohno and S. Hyoda: J. Phys. D 7, 1243 (1974); D. Maugis, G. Desalos-Andarelli, A. Heurtel, and R. Courtel: ASLE Transaction 21, 1 (1977); M. D. Pashley, J. B. Pethica, and D. Tabor: Wear 100, 7 (1984)
2. R. J. Walko: Surface Sci. 70, 302 (1978); D. H. Buckley: Surface Effects in Adhesion, Friction, Wear and Lubrication, (Elsevier, Amsterdam, 1981)
3. G. Binnig, C. F. Quate, and Ch. Gerber: Phys. Rev. Lett. 56, 930 (1986)
4. C. M. Mate, G. M. McClelland, R. Erlandsson, and S. Chiang: Phys. Rev. Lett. 59, 1942 (1987)
5. R. Erlandsson, G. Hadziioannou, C. M. Mate, G. M. McClelland, and S. Chiang: J. Chem. Phys. 89, 5190 (1988)
6. J. N. Israelachvili, P. M. McGuiggan, and A. M. Homola: Science 240, 189 (1988); A. M. Homola, J. N. Israelachvili, M. L. Gee, and P. M. McGuiggan: J. Tribology, to be published
7. A. I. Bailey and J. S. Courtney-Pratt: Proc. R. Soc. London A 227, 500 (1955); J. N. Israelachvili, and D. Tabor: Nature (London) Phys. Sci. 241, 148 (1973); J. Israelachvili and D. Tabor: Wear 24, 386 (1973); B. J. Briscoe and D. C. B. Evans: Proc. R. Soc. London A 380, 389 (1982)
8. P. M. McGuiggan, J. N. Israelachvili, M. L. Gee, and A. M. Homola: in New Materials Approaches to Tribology: Theory and Applications, Eds. L. E. Pope, L. Fehrenbacher, and W. O, Winer (MRS, Boston, 1989), to be published
9. J. N. Israelachvili, private communication
10. J. van Alsten and S. Granick: Phys. Rev. Lett. 61, 2570 (1988); Tribology Trans., to be published.

11. U. Landman, W. D. Luedtke, and M. W. Ribarsky: in <u>New Materials Approaches to Tribology: Theory and Applications,</u> Eds. L. E. Pope, L. Fehrenbacher, and W. O, Winer (MRS, Boston, 1989). to be published; M. W. Ribarsky and U. Landman: Phys. Rev. B <u>38,</u> 9522 (1988)

12. W. A. Goddard: private communication

13. I. P. Batra and S. Ciraci: J. Vac. Sci Technol. A <u>6,</u> 313 (1988); F. F. Abraham, I. P. Batra, and S. Ciraci: Phys. Rev. Lett. <u>60,</u> 1314 (1988); F. F. Abraham and I. P. Batra: Surf. Sci. Lett. (in press)

14. J. Ferrante and S. V. Pepper: ASLE Transactions <u>21,</u> 14 (1977); N. Gane, ASLE Transactions <u>21,</u> 15 (1977); N. Gane and J. Skinner: Wear <u>25,</u> 381 (1973)

15. see, for example: G. Binnig, Ch. Gerber, I. Stoll, T. R. Albrecht and C. F Quate: Europhys. Lett. <u>3,</u> 1281 (1987); T. R. Albrecht and C. F. Quate: J. Vac. Sci. Technol. A <u>6,</u> 271 (1988); O. Marti, B. Drake, S. Gould, and P. K. Hansma: J. Vac. Sci. Technol. A <u>6,</u> 287 (1988); O. Marti, H. O. Ribi, B. Drake, T. R. Albrecht, C. F. Quate, and P. K. Hansma: Science <u>239,</u> 50 (1988); S. Gould, O. Marti, B. Drake, L. Hellemans, C. E. Bracker, P. K. Hansma, N. L. Keder, M. M. Eddy, and G. D. Stucky: Nature <u>332,</u> 332 (1988)

16. for a review, see P. Bak: Rep. Prog. Phys. <u>45,</u> 587 (1982)

17. for a review, see G. Grüner and A. Zettl: Phys. Repts. <u>119,</u> 117 (1985)

18. for a review, see W. Dieterich, P. Fulde, and I. Peschel: Adv. in Phys. <u>29,</u> 527 (1980)

19. for a review, see J. H. van der Merwe, J. Woltersdorf, and W. A. Jesser, Matls. Sci. and Eng. <u>81,</u> 1 (1986)

20. M. H. Grabow and G. H. Gilmer: Mat. Res. Soc. Symp. Proc. <u>56,</u> 13 (1986)

21. Y. I. Frenkel and T. Kontorowa: Zh. Eksp. Teor. Fiz. <u>8,</u> 1340 (1938)

22. F. C. Frank and J. H. van der Mewe, Proc. R. Soc. <u>198,</u> 205, 216 (1949)

23. S. Aubry: Ferroelectrics <u>24,</u> 53 (1980); M. Peyrard and S. Aubry: J. Phys. C16, 1593 (1983)

24. J. B. Sokoloff: Surface Sci. <u>144,</u> 267 (1984)

25. G. A. Tomlinson, Phil. Mag. S. <u>7,</u> Vol. 7, 905 (1929)

26. H. Goldstein: <u>Classical Mechanics,</u> 2nd. ed. (Addison-Wesley, Reading, 1980).

27. H. Reiss: J. Appl. Phys. <u>39,</u> 5045 (1968)

28. for a review, see D. Kashchiev: Surface Sci. <u>86,</u> 14 (1979)

29. M. J. Sutcliffe, S. R. Taylor, and A. Cameron: Wear <u>51,</u> 181 (1978)

Structural and Dynamical Properties of Langmuir-Blodgett Crystals

C. Wöll[1,*] *and V. Vogel*[2]

[1]IBM Almaden Research Center, 650 Harry Road, San Jose, CA 95120, USA
[2]University of California, Department of Physics, Berkeley, CA 94720, USA
*Permanent address: Institut für Angewandte Physikalische Chemie,
 Universität Heidelberg, Im Neuenheimer Feld 253,
 D-6900 Heidelberg, Fed. Rep. of Germany

Langmuir-Blodgett (LB)-films of fatty acids and their salts exhibit a crystalline structure when properly prepared at the air-water interface. This has been known for a long time, but mainly from indirect evidence. Only recently, with the refinement in experimental techniques, has a direct determination of the spatial order within the LB-films become possible on the water surface, as well as after deposition on solid substrates. All the experimental data which are available so far indicate structural properties, i.e. coherence length, defect concentration, long range correlation, which justify the term "Langmuir-Blodgett crystals". However, there are some substantial differences to usual crystals, which show up most strongly in the dynamical properties of these films. Recent key experiments which now provide new insights into the structural and dynamical properties of LB-crystals will be reviewed, with a special emphasis on films made out of arachidic acid ($C_{19}H_{39}COOH$) and its salts.

Introduction

With Langmuir and Blodgett's observation [1] in 1937 that monolayers of long chain carboxylic acids can be transferred from a water surface to a solid surface by sequentially dipping the substrate through the monolayer covered air/water interface, a widespread research field was opened. Superlattices of organic molecules with up to several hundred monolayers of varying chemical composition can be fabricated by the use of this "Langmuir-Blodgett" (LB) technique. Conventionally the first step for assembling LB-films from appropriate substances, e.g. fatty acids with chain lengths between 16 and 22 C-atoms, consists of preparing a stable film at the air-water interface, where all chain molecules are aligned approximately normal to the interface. By the LB-technique these films can be deposited on appropriate substrates which will preserve the monolayer structure and allows manufacture of stacks of monolayers up to, say, a 1000 layers.

Systems of stacked monolayers are of interest for a large variety of applications [2], they can serve as a matrix to embed molecules with functional units, such as dyes and proteins and they can be used as spacer layers, since their layer thickness is given by the chain length of the aliphatic molecule and the tilt angle of the molecular axis with respect to the surface normal. For arachidic acid the layer thickness amounts to 25 Å [3]. In additon fatty acid monolayers are transparent in the visible and provide a low dielectric constant.

Structural imperfections within the multilayer systems can give rise, however, to problems in a variety of applications, such as the development of electronic devices using

LB-films as an insulator [4]. The need of a thorough understanding of the structural properties of pure carboxylic acid monolayers at the air-water interface, as well as for deposited layers is clear. With the help of new surface sensitive methods and the refinement of older techniques during the last years considerable progress towards an understanding of the structure of LB-films on a molecular basis has now been achieved.

For those systems where the structure is sufficiently well known, it is interesting to investigate the dynamics, i.e. the spectrum of intramolecular and intermolecular vibrational excitations. Although the vibrations at higher energies (e.g. C-H stretch vibrations) have been investigated in some detail there is little experimental or theoretical information available for vibrations at thermal energies, such as elastic waves, within deposited monolayers. In general an analysis of the dynamical properties of a system can substantially corroborate structural information and reveal details on the interaction between molecules, which may not be obtainable with structurally sensitive techniques alone. As it will be shown later, especially in case of the LB-crystals, there are certain structural peculiarities which might be related to specific dynamic anomalies within these films.

We will confine the scope of this article to a rather small class of materials, namely arachidic acid and some of its salts, the basic properties of which will be given in the next part of the paper. Results obtained for this prototype class of materials can then serve as a guide to discuss properties of LB-films prepared from other substances. The number of materials which are suited for preparing oriented monolayers at the air/water interface has increased tremendously in the past years and the reader is referred to recent review articles [5,6,7] for a discussion of the various substances and the properties of monolayers made out of them. After a brief introduction to the preparation of monolayers at the air/water interface we will proceed with a discussion of the basic structural and dynamical properties at the air-water interface. This discussion will be based on a small number of direct experimental results obtained by rather different techniques. The last part will then be devoted to a corresponding analysis of LB-films, which are prepared by repeated depositions of the monolayers on solid substrates.

Bulk structures

The thermodynamically most stable molecular structure of arachidic acid is depicted in Fig 1. All carbon atoms lie within one plane and are arranged in a trans-configuration, the structure lowest in energy for a saturated hydrocarbon chain. Gauche conformations are produced by a 60° rotation of a part of the molecule around a single C-C bond, which will bend the hydrocarbon chain. These gauche structures, which have an excess energy of about 0.8 kcal/mol [8], may serve as an intermediate for disorder transitions into various isomeric forms of chain kinks. A kink isomerisation will consist of two spatially separated gauche conformations with an energy of formation of about 6 kcal/mol [9]. Because of the nonzero activation energy the contribution of the gauche bonds vanishes at low temperatures as confirmed by x-ray structure analyses on single crystals of fatty acids and n-alkanes. However, crystallographic evidence has been found for kinked chain structures in hexatriacontan ($C_{36}H_{74}$) for temperatures as low as 345 K [10].

In general the fatty acids crystallize in sheets [11,12], with all the molecular axes parallel within one sheet but with a nonzero tilt-angle to the layer-normal, similar to the behaviour of the n-alkanes. However, with regard to the structure within the sheets, the

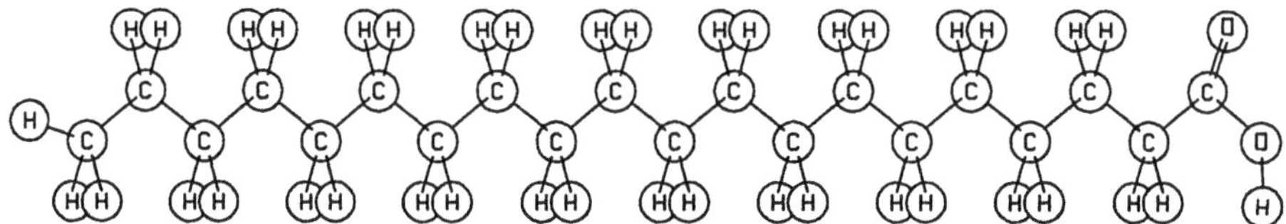

Figure 1: Molecular structure of arachidic acid. All carbon-carbon bonds are in the trans-configuration, the thermodynamically most stable structure.

long chain fatty acids show a polymorphism, and for even-numbered fatty acids three different polymorphs, A,B and C have been observed. Form A exhibits a triclinic crystal structure with a triclinic subcell, whereas form B and C belong to the monoclinic class with a orthorhombic subcell. There is no qualitative difference between forms B and C, both belonging to the same crystallographic class. Form C has a larger tilt-angle of the molecules within the plane, resulting in a smaller interlayer spacing.

Although arachidic acid together with its salts are the most intensively studied materials used to manufacture LB-films, no complete x-ray structure determination for single crystals made out of these materials is available. Only the interlayer distance has been determined [13]. As the homologous series of even-membered fatty acids and their salts have very similar physical and chemical properties, e.g. their melting points show a smooth, monotonic dependency on the number of carbon atoms, the structure of arachidic acid is expected to be very similar to that of stearic acid. The crystal structure of form C of the different polymorphs of stearic acid ($C_{18}H_{36}O_2$), which have been studied by several groups [14,15], is given in Fig 2. The rare A form of stearic acid has not been analyzed crystallographically, but the structure is expected to be similar to that of the A-form for dodecanoic acid, which has been determined [16]. The fact that only the metastable A form of the fatty acids shows a subcell analogous to that of simple, even-numbered hydrocarbon chains, namely only one molecule in a triclinic unit cell, is indicative of the strong structural influence of the carboxylic headgroups. As is apparent from Fig.2, the interaction between the hydroxylic groups dominates the interaction between the hydrocarbon tails and favours an orthorhombic unit-cell, where adjacent molecules are rotated along their axis with respect to each other. For stearic acid the melting temperature is 67.7 °C, and a irreversible transition from form B to form C occurs at 46 °C.

The structural properties of the fatty acid salts have been less extensively investigated. Crystallographic studies indicate that the heavy metal salts of stearic acid and palmitic acid [17] exhibit a lamellar structure, consisting of two-dimensional counter-ion layers, which are separated by a double layer of hydrocarbon tails. The distance between consecutive sheets generally amounts to slightly less than twice the length of the molecule. Similar to the case of the fatty acids a polymorphism is observed, and three different phases have been identified [18]. Whereas the unit-cell dimensions have been determined for a number of different soaps [18] we are aware of only two complete crystal structure determinations, namely for the form A of K-caprate $C_6H_{11}O_2K$ [19] and for the B-form of K-palmitate $C_{16}H_{31}O_2K$ [20] .

An x-ray investigation on Cd-arachidate [21] revealed three different phases. But only the low-temperature phase produced a diffraction pattern indicative of a well ordered crystalline structure. For the other two phases, which exist at temperatures above 110 °C,

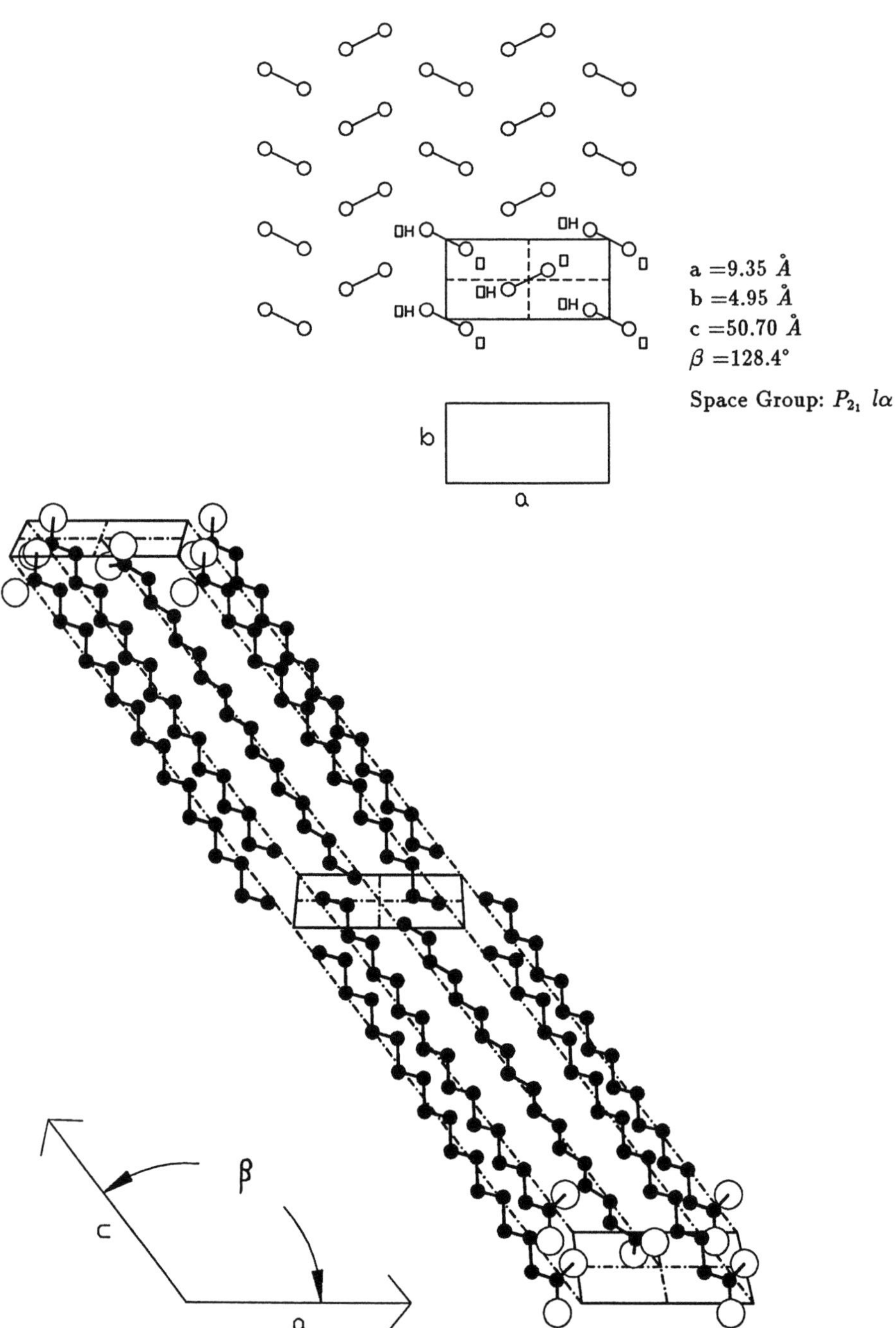

Figure 2: Crystal structure of form C of stearic acid. In the upper part the crystallographic data (from Ref. 15) is displayed together with the orthorhombic two-dimensional subcell. In the lower part a three-dimensional projection of the complete monoclinic unit cell is shown.

considerable disorder was observed, together with a decrease of the interlayer distance. The onset of disorder was again accompanied by a change of the hydrocarbon-subcell to an hexagonal one. In general the degree of disorder within single crystals of fatty acid soaps is considerably higher than for the pure fatty acids. The strong tendency towards disorder has been explained by the formation of stacking faults and random rotations of the fatty acid tails around their molecular axes [22].

Preparation and structure at the air-water interface

The standard procedure [3,23] for preparing arachidic acid monofilms at the air-water interface consists of spreading a defined volume of arachidic acid molecules dissolved in an organic solvent (typically $CHCl_3$) on the surface of high-purity water contained in a trough which is equipped with a moveable barrier. After the solvent has evaporated the molecules are subject to competing forces: because of the interaction between the hydrophobic fatty acid tails there is a tendency of the molecules to form 3-dimensional aggregates. On the other hand the strong interaction of the hydrophilic headgroups with the water molecules favours a solvation of the molecule into the water. The former effect dominates for chains with more than about 25 carbon atoms, the latter for chain lengths shorter than 12. For fatty acids with chains of intermediate length a compensation occurs and the formation of a stable, two dimensional monolayer at the air/water interface becomes possible.

The film at the water surface can now be compressed by a moveable barrier while the surface pressure is monitored by an appropriate device [3,23]. A typical compression isotherm for arachidic acid is given in Fig 3, where the surface pressure is plotted against the area per molecule. For large areas per molecule (the bulk value is 20 $\mathring{A}^2$) the surface pressure is close to zero (large compressibility). With fluorescence microscopy [24] it could be demonstrated that for stearic acid this phase consists of 2-dimensional islands of up to 100 μm in diameter. There is strong indication that within these islands the molecules lie flat on the water surface. Further compression of the film closes the gap between islands and leads to the formation of a continuous film at about 35-40 $\mathring{A}$/molecule. Noticeably this phase transition is not accompanied by any observable change in the compressibility, the surface pressure stays below 0.1 mN/m (see Fig 3). This observation is consistent with the assumption, that the angle between surface normal and the chains of the fatty acid, which is 90° for large areas per molecule (hydrocarbon chain lying flat on water surface), continuously decreases to 0° after the formation of a continuous film.

But upon continuing to reduce the area per molecule in the film further, a steep increase in the surface pressure can be observed at 24 $\mathring{A}^2$/molecule and a second kink at about 20 $\mathring{A}^2$/molecule. At this point the compressibility of the monolayer equals that of the bulk. The area per molecule at this second kink shows a dependence on temperature but not on the chain length when using different fatty acids. From this fundamental observation Langmuir [25] could conclude in 1917 that amphiphilic molecules orient at the air-water interface with the close-packed chains perpendicular to the interface.

For the fatty acid salts the situation is somewhat more complex, as the counterions have also to be taken into account. However, the experimentally obtained compression isotherms are very similar to those of the corresponding acids. X-ray and neutron reflectivity measurements [26] have shown that for Cd-arachidate the Cd^{2+} ions are tightly bound to the acid head group if the pH of the subphase is greater than 5. Very recently

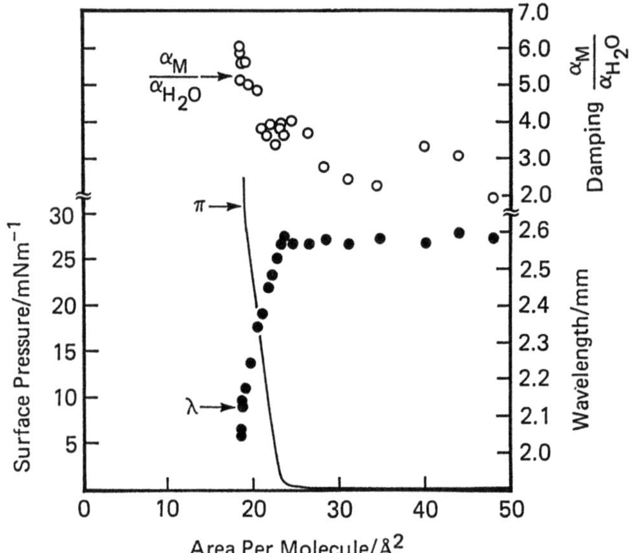

Figure 3: The lower graph (solid line) represents a compression isotherm for docosanoic acid spread on a water surface. The large dots indicate the wavelength of capillary surface waves at 180 Hz, which was measured simultaneously. Note the scale for the wavelengths on the right side. The crosses indicate the damping of the surface waves with respect to the damping of waves with the same frequency on the clean water surface.

similar findings were reported by Bloch et al. [27] for Mn-stearate. In their synchroton x-ray study they could demonstrate in addition, that the Mn-ions show order only for the compressed phase, but not for the expanded phase. As the Mn-ions are bound to the carboxylic end-groups of the fatty acid, this implies that the fatty acid molecules are also not ordered for large areas per molecule, consistent with the above discussion.

Although from the properties of the fatty-acid monolayers as discussed above a closed packed hexagonal arrangement of the fatty-acids seems most likely, a direct confirmation was not possible until recently, when grazing incidence x-ray scattering could be used directly to determine structure and lattice constant within these films [28,29]. The diffraction patterns observed in these measurements turned out to be similar to Debye-Scherrer diffraction patterns, which are obtained when x-rays are scattered from powders. As depicted schematically in Fig 4, Debye-Scherrer diffraction patterns consist of rings and are indicative of small, randomly oriented crystallites. In the measurements of Dutta et al. [28] on Pb-stearate the k-vector of the first diffraction ring amounts to 1.60 Å^{-1} with a halfwidth of about 0.02 Å^{-1}. From that a mean intermolecular distance of 4.53 Å and an average size of the 2d-crystallites of about 250 Å can be concluded. An orthorhombic subcell as expected from the bulk crystal structure of lead-stearate [30] or that of fatty acids for the structure within the lead-stearate monolayer at the water surface can be ruled out, since in that case the two different nearest neighbour distances should give rise to two closely spaced diffraction rings. Instead the data is indicative of a hexagonal packing. It should be noted however, that a hexagonal packing is not compatible with the twofold symmetry of the fatty acids (the backplane of the carbon

22

atoms is a symmetry plane), so that a rotational disorder has to be assumed. These phases are also called hexatic [31], they show only translational ordering (the axes of the molecules are ordered) but no rotational order (there is no long-range correlation between the azimuthal orientation of the backplanes of nearest neighbours).

It is interesting to compare this observation with the properties of long chain alkanes [32]. At low temperatures the molecules crystallize in an orthorhombic subcell and all are rotationally aligned. Increasing the temperature induces a phase transition to a structure with an apparent random distribution of the angles between the plane spanned by the backbone carbon atoms of a single molecule. The appearance of rotational disorder is accompanied by a change of the crystallographic structure to one with a hexagonal subcell.

Dynamical properties of LB-films at the air/water interface

Together with the strong structural changes of the fatty acid monolayer upon compression, the dynamical properties of the film also change noticeably. This is demonstrated for a monolayer of dodecanoic acid by the dependence of the wavelength of 160 Hz capillary surface waves on the area per molecule [33], as shown in Fig 4. The wave is excited by an electrocapillar perturbation of the interface. The wavelength is measured by scanning the surface with a He-Ne laser and monitoring the deflection of the specular beam [34]. Whereas for large areas per fatty acid molecule, corresponding

diffraction pattern

Figure 4: Schematic diffraction patterns for x-rays or electrons scattered from two- dimensional crystallites. If all the crystallites have the same structure, but are rotated randomly with respect to each other, as shown in the upper left, the diffraction patterns of the single crystallites overlap, resulting in Debye-Scherrer rings (upper right). The half-width δk of the Debye-Scherrer rings is a measure of the average diameter of the crystallites. If the crystallites are not rotated with respect to each other, a distinct diffraction picture emerges as shown in Fig 6.

to a small surface pressure, a wavelength typical for capillary waves on a clean water surface is observed, compression leads to a strong wavelength decrease, corresponding to an increase of the restoring forces for distortions perpendicular to the surface. Fig 4 also reveals that the strong change in wavelength is clearly correlated with the first kink in the surface pressure vs. area/molecule isotherm (see last section), which is plotted at the bottom of Fig 4. From the dependence of the damping of the 160 Hz capillary waves on the area per molecule, as shown in the upper part of Fig 4, it is demonstrated that the presence of the monolayer not only enlarges the restoring forces perpendicular to the surface, but also increases the viscosity of the surface, leading to a much stronger damping of the capillary waves.

In addition to the capillary surface waves other types of polarization can be investigated. So far the observation of three different modes of surface waves on monolayer coated air/water interfaces has been reported. (1) Transverse capillary waves [35,36] have a polarization ellipse lying in the plane spanned by the wave-vector and the surface normal, with the larger half-axis parallel to the surface normal. (2) Transverse shear waves [37-41] have the polarization ellipse lying in the interface plane with the larger half-axis perpendicular to the wave vector. (3) In longitudinal or Marangoni waves [42] the polarization ellipse is oriented as that for the transverse capillary waves, but with the larger half-axis along the wave vector.

In recent years specific experimental techniques have been developed to study each of these modes. For the investigation of transverse capillary waves two different techniques are available. With single mode waves generated by electrocapillar deflection of the water surface [43], waves in the frequency range between 100 Hz and 1 kHz could be investigated. An alternative approach is the analysis of oscillations in the intensity of light reflected from the thermally roughened surface [44-47], which covers a frequency range from 1 kHz to 10 kHz. Transverse waves of shear character are usually generated by a rotor that touches the surface. A variety of techniques is available for the generation of surface waves with longitudinal character, e.g. oscillating bars or pressure pulses induced by photoisomerisation of dye molecules embedded in the monolayer [48] by taking advantage of the inherent coupling of transverse capillary and longitudinal waves [49].

A remarkable feature of longitudinal waves in fatty acid monolayers on a water surface is their extremely small sound velocity. Suzuki et al. [48] have employed the photoisomerisation of spiropyran molecules embedded in a lipid monolayer to excite short pressure pulses. They could experimentally determine the pulse velocity for longitudinal waves to 2.6 m/s, which is several orders of magnitude lower than that of water (3000 m/s). Note that the velocity of sound for capillary surface waves (transversely polarized, see discussion above) increases by covering the surface with the fatty acid monolayer.

Unfortunately a direct correlation between these changes of the dynamical properties of the LB-film and the interaction between the fatty acid molecules at the air water interface is difficult to reveal, as the dispersion and in particular the damping of viscoelastic waves at the LB-film covered water surface is strongly influenced by the fluidity of the liquid. Any periodic motion at the air-water interface will be accompanied by a periodic flow in the liquid underneath, which makes a description in the framework of hydrodynamics necessary. Because of this rather complex framework most of the dispersion equations for the various surface waves have until now been derived by treating the monolayer as an isotropic and viscoelastic continuous film (see e.g. Ref.36). Only very few approaches have been reported, where the surface structure is explicitly taken into account [50].

Whereas the presence of counterions in the aqueous subphase does not show a pronounced effect on the structure of compressed monolayers at the air-water interface, the dynamical properties of the monofilms can be strongly affected. However, the pH of the subphase has to be larger than the pK of the fatty acid in order to confine the counterions to a thin layer between monolayer and the water. If this is the case, polyvalent counterions will lead to a bridging of the fatty acid molecules. The most direct way to determine this stiffening quantitatively is to measure the in-plane shear viscosity of the monolayer coated water surface. Indeed strong changes are observed and a strong increase in stiffness is found for divalent counterions [40,51,52] like Mg^{2+}, Ca^{2+} and Cd^{2+}. The trivalent Al^{3+} ion gives rise to even larger shear moduli of up to 410 mN/m for a close-packed arachidate monolayer [40].

Structural properties of deposited films

The possibility to deposit the monolayers after preparation on the air/water interface on solid substrates while retaining the microscopic molecular arrangement (Langmuir-Blodgett technique) is the basis for nearly all applications of LB-films. In Fig 5 the procedure for deposition on hydrophilic materials is shown, where the substrate is immersed in the water before preparation of the monolayer. The only difference for hydrophobic substrates (like graphite) is the omission of step 1 in Fig 5, resulting in a deposited film with the hydrocarbon tails being in contact with the surface. After deposition of the fatty acid monolayer the substrates can be transferred to vacuum; the deposited LB-films are quite stable and are even compatible with UHV. Once in vacuum, a large variety of techniques for structural investigations can be applied, especially electron diffraction, which is the standard techniques for structure determinations on clean and adsorbate covered solid substrates. The potential of applying Low Energy Electron Diffraction (LEED) to the study of LB-films was realized very early [53]. Fig 6 shows a LEED-pattern obtained from a mixture of arachidic acid and Methyl Stearate (ratio 9:1) deposited on a single crystal (111)-surface of copper. In contrast to the Debye-Scherrer rings in the x-ray diffraction pattern for the LB-film at the water surface a distinct, hexagonal array of diffraction spots is observed, indicating that the 2D-crystallites at the air-water interface have been oriented during deposition (the electron beam diameter is about 0.25 mm). A remarkable consequence of these LEED experiments [54] is that the orientation of the LEED-diffraction pattern from the monolayer did not change when the electron beam was scanned over the LB-covered area (25 mm) of the substrate, a noble metal (Cu, Ag and Au) single crystal with a (111) surface orientation, epitaxially grown on mica. This observation indicates that during the monolayer deposition an alignment of the randomly oriented crystallites at the air/water interface has occurred. Again the apparent hexagonal order raises the question of whether the subcell is orthorhombic or truly hexagonal, the latter implying a rotational disorder (hexatic phase). Similar results to those displayed in Fig 6 have also been obtained for deposited films of Cd-stearate [55], where a coherence length of several mm was observed.

Although it is quite straightforward to extract the translational order and the lattice constant of the deposited LB-films from these the electron diffraction patterns, it is much more difficult to obtain information on the tilt-angle between surface normal and the fatty acid molecule axis. However, a direct determination of this tilt-angle for single deposited films has become possible by the recent development of the synchroton-radiation based technique NEXAFS (near edge x-ray absorption fine structure). Outka et al. [56] have

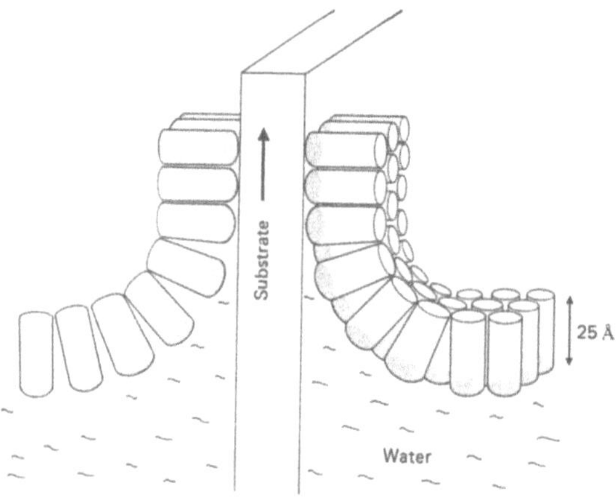

Figure 5: Transfer of fatty-acid monolayers from air-water interface onto solid substrates (Langmuir-Blodgett technique). The fatty acid molecules are represented by cylinders, with the black end denoting the hydrophilic (carboxylic) head-group. The picture shows deposition for the case of a hydrophilic substrate (e.g.clean metal surfaces), where the first layer is deposited by withdrawing the substrate out of the water through the monolayer covered air/water interface. For a hydrophobic substrate (e.g. graphite) the process is reversed, the first layer is deposited by dipping the substrate into the water. The fabrication of multilayers can be accomplished by repeatedly withdrawing and dipping the substrate.

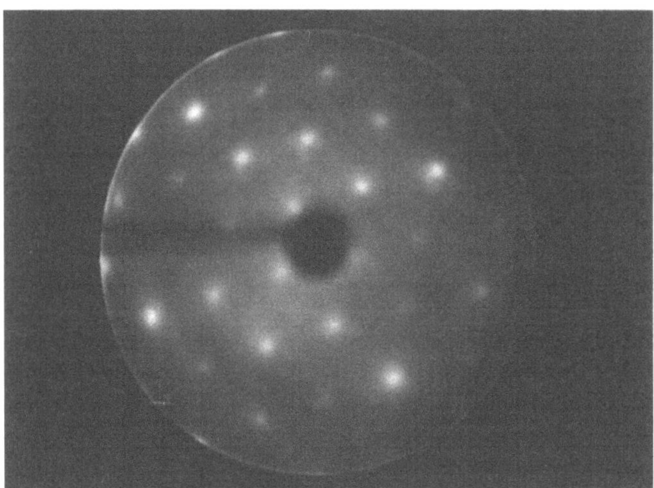

Figure 6: LEED-pattern (104 eV) of a single monolayer of arachidic acid and methyl stearate deposited on a single crystal Cu(111) surface. The diffraction pattern is interpreted as revealing the formation of a structured monolayer crystal with a lattice constant of 4.9 $\mathring{A}$ and a coherence length of more than 140 $\mathring{A}$. For further details see Ref. 54

performed measurements on arachidic acid, Cd-arachidate and Ca-arachidate deposited on oxidized Si(111). From their data they conclude that for Cd-arachidate the molecular axis is parallel to the surface normal, whereas Ca-arachidate exhibits a tilt-angle of 33° This tilt-angle is also observed in the bulk of fatty acids. No ordering was observed for the arachidic acid films. The electronic states of LB-films made from cadmium arachidate have been investigated with UPS (ultraviolet photoelectron spectroscopy) [57] and a good agreement with theoretical calculations has been obtained.

Several attempts have been made to obtain structural information on deposited LB-films by using an STM (scanning tunneling microscope). STM-images for bi-layers of cadmium arachidate [58,59,60] on graphite were interpreted as an evidence of an ordered array of aliphatic chains with hexagonal symmetry. However, the electric conductance of the aliphatic chains as observed in the STM investigations is at odds with the values obtained in a thorough investigation by Mann et al. [61] (see also Ref.62). The discrepancy amounts to several orders of magnitude and it is not yet understood to what precisely the periodic structure observed in the STM-images is related to.

Whereas single deposited layers of fatty acid or fatty acid salt molecules might differ in structure from the corresponding bulk single crystals, increasing the number of deposited layers should eventually show a transition of the ordering within the deposited films to that of the bulk structures. Note, that the preparation of the multilayer systems by the Langmuir-Blodgett technique as described above (y-deposition, see Fig 5) results in an headgroup-to-headgroup and tails-to-tails arrangement of molecules, which is identical to the structure of the bulk-phases (see Fig 2). For a number of systems, such as Pb-stearate [30] and Ca-stearate [63], a close agreement between the structure of multilayer films and of bulk single crystals has been reported.

The most prominent feature of deposited multilayer systems is the lamellar structure, and a variety of techniques (e.g. x-ray scattering [13], neutron scattering [64,65]) have been used to determine the interlayer distances. These experiments indicate that very few faults are present as far as the stacking of layers is concerned, and very close agreement between the interlayer distance for multilayer systems and bulk crystals from the same material was found.

However, from Normaski interference microscopy [66], from TEM (Transmission Electron Microscopy) and REM(Reflection Electron Microscopy) [66,67] studies it is well known that pinholes and fractures exist within mono- and multilayers for all substrates. The amount and size of these defects depends on a variety of factors, such as deposition procedures and structural and chemical properties of the substrate, and represents the major problem in connection with technical applications of LB-films. However, many experimental results indicate that on the microscopic scale a high degree of order is present.

For the intralayer structure a gradual transition from the commonly observed hexagonal arrangement within a single deposited monolayer to a (bulk) orthorhombic subcell in multilayer systems has to be expected. Indeed this has been observed for docosanoic acid deposited on thin films of carbon and aluminum in a combined RHEED (Reflection High Energy Electron Diffraction) and IRS study [68]. The same observation has been reported for LB-films of stearic acid with between 1 and 9 layers deposited on germanium [69]. From these IR-experiments the authors conclude that the the transition from the hexagonal or quasi-hexagonal ordering in the first layer to the orthorhomic structure present in the bulk starts as early as in the second layer. Raman spectroscopy has also been employed [70] to study this interesting change of structure. For the case of Cd-stearate deposited on fused silica substrates multilayer systems with between 1 and 27 layers were investigated. The results again indicate a hexagonal packing within the first few layers whereas for more than about 5 layers a transition to an orthorhombic packing is reported. Not surprisingly this change in crystal structure is accompanied by a significant increase in lattice defects. Considerable disorder in the first deposited films, apparently linked to this structural misfit was also observed in a FTIR study of Cd-arachidate deposited on gold by Arndt et al. [71].

Dynamical properties of deposited films

Due to the large number of atoms in a fatty acid molecule (60 for arachidic acid) the dynamics within the deposited films becomes rather complicated. The possible vibrations occur in two different energy regions, one from about 40 cm^{-1} up to 3200 cm^{-1} containing the intramolecular vibrations ($3x60$-$6 = 174$ different normal modes are expected for arachidic acid), and the other one below 40 cm $^{-1}$, containing the intermolecular vibrations.

Intramolecular modes within hydrocarbon chains have been extensively studied for the alkanes and there is a very satisfying agreement between experimental data and normal mode analyses [72,73]. The ability to record and analyze experimental data for the vibrations within these films has reached the point where even subtle features can be understood. For example a splitting of about 20 cm^{-1} for some of the rocking modes in alkanes could be traced back experimentally as well as theoretically, to a crystal field splitting [74]. The splitting occurs because in crystal structures with more than one atom per unit cell all modes occur in multiples (optical modes). For the case of an orthorhombic subcell with two atoms per unit cell two branches, one acoustic and one optical, result from the splitting. The magnitude of the splitting depends on the polarization vector of the normal mode and on the strength of the intermolecular interaction. Rocking modes turn out to be very sensitive to this effect. We give this example in order to demonstrate that the understanding of intramolecular vibrations has reached the point where a refined analysis of spectroscopic data can reveal details on structural properties too.

The standard technique for obtaining vibrational spectra is infrared spectroscopy, which may be applied in a number of modifications (Fourier Transform Infrared Spectroscopy (FTIR), Infrared Reflection Absorption Spectroscopy (IRAS)). Recently Raman Spectroscopy has also been applied to obtain vibrational spectra for LB-films from fatty acids. Infrared Spectroscopy and Raman spectroscopy are partially complimentary to each other, because among the possible normal modes of a molecule only a subset is infrared active, so that the number of vibrational frequencies, which can be determined experimentally, can be considerably increased by additionally obtaining information on the Raman active modes. One particularly interesting point in connection with both Infrared Spectroscopy and Raman spectroscopy is the possibility of obtaining polarized spectra, which can deliver important information on the orientation of the molecules.

In one of the first applications of IRS to study structural properties of LB-films Rabolt et al. [75] have studied deposited films of Cd-arachidate. When recording polarized IR-spectra for electric field vectors parallel and perpendicular for the substrate, they observed strongly increased excitation probabilities for the CH_2 bending modes for the parallel polarized case, which allowed them to conclude that the hydrocarbon chains of the

Cd-arachidate LB-systems were oriented with their molecular axis parallel to the surface normal. In addition Rabolt et al. [75] were also able to observe a splitting of a rocking mode (see discussion above), which allowed them to conclude that the crystallographic structure of the deposited monofilms exhibits an orthorhombic subcell. Together with the information that the molecular axes are oriented parallel to the surface normal they were able to propose an orthorhombic crystal structure with four molecules per unit cell. It is interesting to note that this structure is very similar to the bulk structure of both the B and C forms of fatty acids, the main difference being that the tilt-angle of the molecular axes within one "layer" is reduced to 0° for the deposited monolayers, thus transforming the bulk monoclinic unit cell into a orthorhombic one.

In a further study Naselli et al. [76] applied the unique sensitivity of the IRS-excitation probability of certain modes to the orientation of the molecular axis in order to study the melting of the deposited films. By comparing relative intensities for vibrations localized in the hydrocarbon tails and vibrations localized in the head-groups they were able to show that the melting occurs in a two stage process. The first step apparently consists of a strong increasing disorder in the arrangement of the hydrocarbon tails at temperatures well below the bulk melting point, whereas a breaking of the head-group lattice occurs at about 100°C, close to the bulk melting point. The first transition was reported to be reversible and might fall in the class of second order order-disorder transitions.

Intermolecular vibrations in deposited LB-films

In addition to the intramolecular vibrations, collective vibrations of intermolecular character are expected. Elastic waves in LB-films have already been discussed above, when the dependence of the wavelength of a capillary wave in a LB-film at the air-water interface was demonstrated. However, in that case the elastic properties of the floating LB-film are considerably influenced by the strong interaction between the hydrophilic ends of the fatty acid molecules and the water, so that the properties of the air-water interface will be dominant and veil the influence of the LB-film itself. When the monolayers are deposited on a solid substrate the vibrations of transverse character within the LB-film are more strongly decoupled from those of the substrate, as the restoring forces for shear displacements in the substrate are now very large. In the case of the liquid (water) surface, the very weak restoring forces to shear displacements arise from the surface tension (within a liquid there are no restoring forces for shear displacements).

The elastic properties of thin layers on solid substrates have been investigated theoretically by Farnell and Adler [77]. In our case the elastic constant of the thin layer (the LB-film) is considerably weaker than that of the substrate, so that we expect the vibrations lowest in energy, the Rayleigh mode and the Love-mode, to be mainly localized in the thin layer, the penetration into the substrate decreasing with decreasing wavelength (the penetration depth of surface localized modes is roughly proportional to the wavelength). An experimental determination of the dispersion curves for the Rayleigh and the Love-mode would offer a unique possibility to study the elastic properties of deposited LB-films, especially as the Rayleigh-mode is a transverse wave, being mainly sensitive to force-constants parallel to the surface normal, whereas the Love-modes are shear-horizontal waves, being mainly sensitive to force-constants for bonds within the surface. Therefore the expected strong anisotropy within the LB-films should be reflected in a strong difference in the energies for Rayleigh-waves and Love-waves. Unfortunately the energy regime where these modes are expected is difficult to access with the techniques which have so far been used to determine dispersion curves of surface localized vibrations, EELS (Electron Energy Loss Spectroscopy) [78] and HAS (Helium Atom Scattering) [79]. Although EELS data obtained for deposited LB-films [80] clearly show the intramolecular modes of the fatty acid molecules in agreement with the IRS data, no conclusion could be drawn on the collective vibrations. The scattering of helium and heavier noble gas atoms [81,82] showed a strong interaction with low-energy modes within the film, but no single phonon modes could be resolved. In fact nearly all the scattered intensity was observed to flow into inelastic channels, no elastic reflection or diffraction could be observed [82]. This might not be too surprising, as the velocity of sound determined for LB-films at the air-water interface is about two orders of magnitude lower than that for surface waves on solid substrates. From this one can estimate the corresponding phonon energies to be lower than about 0.2 meV, which is somewhat below the present resolution of the HAS-technique, although recently an improvement of the resolution to 0.1 meV and below has been reported.

There is only one piece of direct experimental evidence for elastic waves in deposited LB-films [83], obtained with Brillouin-scattering, a laser-based technique which has in the past been very successfully used to determine dispersion curves for large wavelengths for surface and interface modes for a variety of materials [84]. In the experiment by Zanoni et al. [83] the Rayleigh mode in LB-films of Cd-arachidate deposited on Mo containing between 11 and 60 monolayers could be determined. From the dispersion of this mode the elastic properties for distortions parallel to the surface normal could be determined, and were found to be similar to those of other organic materials. However, as pointed out above, it is actually the Love-modes which are expected to contain the most direct information on the interaction between the fatty acid molecules, as these waves have a shear horizontal character. Unfortunately these vibrations could not be observed in the Brillouin scattering data. From this the authors concluded that the energy of these modes is smaller than their resolution, which gives an upper limit of 4.0×10^{-8} N/m^2 for the elastic constant c_{44} within the LB-film (parallel to the surface). This value is extremely low compared to other organic materials in the solid state and indicates a strong anisotropy within these films, with normal restoring forces for shear waves polarized perpendicular to the surface (Rayleigh) and very weak restoring forces for shear waves polarized parallel to the surface.

It seems, that the results obtained with HAS [81,82] and those obtained with Brillouin scattering [83] are not completely in agreement. From the Brillouin-scattering data it was concluded that the Rayleigh-mode has a velocity of sound of about 2500m/s, so that the resolution of HAS should in principle be sufficient to resolve the excitation of single phonons, as has been demonstrated for the surfaces of many materials, metals as well as insulators [79]. But even in the high resolution HAS data [82] only broad energy losses down to very low energies were observed, which cannot be related to interactions with shear horizontal modes (such as the Love-mode), as for symmetry reasons the interaction probability is zero. It should be noted, however, that the HAS experiments were done for LB-films consisting of a single monolayer, whereas the Brillouin-scattering (for intensity reasons) was done for LB-films of at least 11 monolayers. Therefore the apparent disagreement could be indicative of a somewhat different microscopic structure of single monolayers, for example the presence of a hexatic phase for the single monolayer could explain the observed behaviour.

Summary

In this article we have given a condensed description of the structural and dynamical properties of Langmuir-Blodgett films made out of arachidic acid and its salts. Although the class of materials suitable for the fabrication of LB-films is much larger, this set of materials has been studied extensively and may serve as a prototype class to discuss the basic physical properties of monolayers and LB-systems. We have related structural parameters of fatty-acid monolayers on the water surface as determined recently by in-situ x-ray scattering and of deposited LB-films as determined by infrared spectroscopy, electron diffraction and near edge x-ray absorption fine spectroscopy to corresponding bulk properties and concluded, that these films exhibit properties that justify the term "Langmuir-Blodgett" crystals. However, there are some peculiarities in connection with the dynamic properties of deposited films which have recently been investigated with He-atom scattering and Brillouin scattering. The recent success in studying fatty acid monolayers on the water-surface and after deposition on solid substrates by in-situ x-ray scattering and near edge x-ray absorption fine structure spectroscopy as well as with infrared spectroscopy has considerably advanced the understanding of structural properties. With the help of infrared spectroscopy and Raman spectroscopy the intramolecular vibrations have been thoroughly characterized and related to structural properties. In addition there is the prospect of applying novel techniques, such as infrared-visible sum-frequency generation [85] to study fast molecular dynamics. The applicability of this method has already been demonstrated for a pentadenanoic acid monolayer on water [86], but there are still a number of unanswered questions concerning the intermolecular vibrations in the thermal and subthermal regime. As this point is of pronounced importance for applications of LB-films further effort is called for ot improve the resolution in both He-atom scattering and Brillouin scattering.

Acknowledgements

We would like to thank Dr.J.Swalen and Dr.V.M.Hallmark for enlightening discussions and careful reading of the manuscript. V.V. thanks the Max-Planck Gesellschaft for a stipend.

References

*Permanent address: Institut für Angewandte Physikalische Chemie, Universität Heidelberg, Im Neuenheimer Feld 253, 6900 Heidelberg, F.R.G

1. K.B.Blodgett and I.Langmuir, *Phys. Rev.* **51**, 964, (1937).

2. see the papers presented at the International Conference on Langmuir-Blodgett Films, in: *Thin Solid Films* **99** (1983); ibid. **132-134**(1985); ibid. **158, 159**(1988).

3. H.Kuhn, D.Möbius and H.Bücher, in: *Physical Methods of Chemistry* edited by A.Weissberger and W.B.Rossiter Wiley, NewYork, 1982, Vol. 1 p.650ff

4. G.G.Roberts, K.P.Prande and W.A.Barlow, *Solid State and Electronic Devices* **2**, 169, (1978).

5. V.K.Agarwal, *Physics Today* **6/1988**, 40, (1988).

6. R.H.Tregold, *Rep. Prog. Phys.* **50**, 1609, (1987).

7. J.D.Swalen, D.L.Allara, J.D.Andrade, E.A.Chandross, S.Garoff, J.Israelachvili, T.J.McCarthy, R.Murray, R.F.Pease, J.F.Rabolt, K.J.Wynne and H.Yu, *Langmuir* **3**, 932, (1987).

8. K.S.Pitzer, *J. Chem. Phys.* **8**, 711, (1940).

9. S.H.Northrup and M.S.Curvin, *J.Phys.Chem.* **88**, 4707, (1985).

10. D.L.Dorset,B.Moss, J.C.Wittmann and B.Lutz, *Proc. Natl. Acad. Sci. USA* **81**, 1913, (1984).

11. E.von Sydow, *Arkiv För Kemi* **9**, 231, (1955).

12. R.M.Clark, *Applied X-Rays* McGraw-Hill, New York, 1955, p.608

13. Piper, Malken and Austin, *J. Chem. Soc.* **129**, 2310, (1926).

14. E.von Sydow, *Acta Cryst.* **8**, 557, (1955).

15. V.Malta, G.Celotti, R.Zannetti and A.Ferrero-Martelli, *J. Chem. Soc.(B)*, 548, (1971).

16. E.von Sydow, *Acta Chem. Scand.* **10**, 1, (1956).

17. R.D.Vold and G.S.Hattiangdi, *Ind. Eng. Chem.* **41**, 2311, (1948).

18. T.R.Lomer, *Acta Cryst.* **5**, 11, (1952).

19. V.Vand, T.R.Lomer and A.Lang, *Acta Cryst.* **2**, 214, (1949).

20. J.H.Dumbleton and T.R.Lomer, *Acta Cryst.* **19**, 301, (1965).

21. P.P.A.Spegt and A.E.Shoulios, *Acta Cryst.* **16**, 301, (1963).

22. A.J.Stosick, *J. Chem. Phys.* **18**, 1035, (1950).

23. G.Gaines, *Insoluble Monolayers at Gas-Liquid Interfaces*, John Wiley and Sons, New York, 1966.

24. M.Moore, C.M.Knobler, D.Broseta and F.Rondelez, *J.Chem.Soc., Faraday Trans.* **2**, 1753, (1986).

25. I.Langmuir, *J. Am. Chem. Soc.* **39**, 1848, (1917).

26. M.J.Grundy, R.M.Richardson, S.J.Roser, J.Penfold and R.C.Ward, *Thin Solid Films* **159**, 43, (1988).

27. J.M.Bloch, W.B.Yun, X.Yang, M.Ramanathan, P.A.Montano and C.Capasso, *Phys. Rev. Lett.* **61**, 2941, (1988).

28. P.Dutta, J.B.Peng, B.Lin, J.B.Ketterson, M.Prakash, P.Georgopoulos and S.Ehrlich, *Phys. Rev. Lett.* **58**, 2228, (1987).

29. K.Kjaer, J.Als-Nielsen, C.A.Helm, L.A.Laxhuber and H.Möhwald, *Phys. Rev. Lett.* **58**, 2224, (1987).

30. J.F.Stephens and C.Tuck-Lee, *J. Appl. Cryst.* **2**, 1, (1969).

31. S.B.Dierker and R.Pindak, *Phys. Rev. Lett.* **59**, 1002, (1987).

32. G.-R. Strobl, *Phys. Bl.* **33**, 550, (1977).

33. V.Vogel, Dissertation, Franktfurt/Main 1987

34. C.H.Sohl, K.Miyano, *Phys. Rev. A* **20**, 616, (1979)

35. R.S.Hansen and J.A.Mann, *J. Applied Physics.* **35**, 152, (1964).

36. E.H.Lucassen-Reynders and J.Lucassen, *Adv. Colloid Interface Sci.* **2**, 347, (1969).

37. J.Lucassen and M. Van den Tempel, *J. Colloid Interface Sci.* **41**, 491, (1972).

38. J.A.De Feijiter, *J. Colloid Interface Sci.* **69**, 375, (1979).

39. H.A.Watermam, *J. Colloid Interface Sci.* **101**, 377, (1984).

40. B.M.Abraham, J.B.Ketterson, K.Miyano and A.Kueny, *J. Chem. Phys.* **75**, 3137, (1981).

41. M.Kawaguchi, M.Sano, Y.-l. Chen, G. Zografi and H.Yu, *Macromolecules* **19**, 2606, (1986).

42. J.Lucassen, *Trans. Faraday Soc.* **548**, 2221, (1968), **548**, 2230, (1968).

43. C.H.Sohl, K.Miyano and J.B.Ketterson, *Rev. Sci. Instr.* **49**, 1464, (1978)

44. J.A.Mann, *Langmuir.* **1**, 10, (1985).

45. F. De Voeght and P.Joos, *J. Colloid Interface Sci.* **98**, 20, (1984).

46. D.Langevin, *J. Colloid Interface Sci.* **80**, 412, (1981).

47. J.A.De Feijiter, *J. Colloid Interface Sci.* **101**, 377, (1984).

48. M.Suzuki, D.Möbius, R.Ahuja, *Thin Solid Films* **138**, 151, (1986).

49. V.Vogel and D.Möbius, *Langmuir,* **5**, 129, (1989).

50. C.F.Tejero, M.J.Rodriguez and M.Baus, *Phys. Rev. A* **29**, 2179, (1984).

51. J.A.Spink, *Trans. Faraday Soc..* **51**, 1154, (1955).

52. R.M.Richardson and M.Buhaenko, *Thin Solid Films* **159**, 171, (1988).

53. L.H.Germer and K.H.Storks, *J. Chem. Phys.* **6**, 280, (1938).

54. V.Vogel and Ch.Wöll, *J. Chem. Phys.* **84**, 5200, (1986).

55. S.Garoff, H.W.Deckman, J.H.Dunsmuir, M.S.Alvarez and J.M.Bloch, *J. Physique* **47**, 701, (1986).

56. D.A.Outka, J.Stöhr, J.P.Rabe and J.D.Swalen, *J. Chem. Phys.* ,**88**, 4076, (1988).

57. K.Seki,N.Ueno,U.O.Karlsson,R.Engelhardt and E.-E.Koch, *Chem. Phys.* **105**, 247, (1986).

58. D.P.E.Smith, A.Bryant, C.F.Quate, J.P.Rabe, Ch.Gerber and J.D.Swalen, *Proc. Natl. Acad. Sci. USA* 34, 969, (1987).

59. H.Fuchs, *Physica Scripta* **38**, 264, (1988).

60. C.A.Lang, H.J.K.Hoerber, W.T.Haensch, W.Heckl and H.Moehwald, *J. Vac. Sci. Technol.* **A 6**, 368, (1988).

61. B.Mann and H.Kuhn, *J. of Applied Physics* **42**, 4398, (1971)

62. E.E.Polymeropoulos, *J. Applied Physics.* **48**, 2404, (1977).

63. G.L.Clark, R.R.Sterrett and P.W.Leppla, *J. Amer. Chem.Soc.* **57**, 330, (1935).

64. R.M.Nicklow, M.Pomerantz and A.Segmüller, *Phys. Rev. B* **23**, 1081, (1981).

65. R.R.Highfield, R.K.Thomas, P.G.Cummins, D.P.Gregory, J.Mingins, J.B.Hayter and O.Schärpf, *Thin Solid Films* **99**, 165, (1983).

66. Vandenyver, *Thin Solid Films* **159**, 243, (1988).

67. A.Fischer and E.Sackmann, *J. Physique* **45**, 517, (1984).

68. A.Bonnerot, P.A.Cholet, H.Frisby and M.Hoclet, *Chem. Physics* **97**, 365, (1985).

69. F.Kimura, J.Umemura and T.Takenaka, *Langmuir* **2**, 96, (1986).

70. S.B.Dierker, C.A.Murray, J.D.Legrange and N.E.Schlotter, *Chem. Phys. Lett.* **137** 453, (1987).

71. T.Arndt and C.Bubeck, *Thin Solid Films* **159**, 443, (1988)

72. R.G.Snyder and J.H.Schachtschneider, *Spectrochim. Acta* **19**, 85, (1963).

73. J.H.Schachtschneider and R.G.Snyder, *Spectrochim. Acta* **19**, 117, (1963).

74. M.Tasumi and T.Shimanouchi, *J. Chem. Phys.* **43**, 1245, (1965).

75. J.R.Rabolt, F.C.Burns, N.E.Schlotter and J.D.Swalen, *J. Chem. Phys.* **78**, 946, (1983).

76. C.Naselli, J.F.Rabolt and J.D.Swalen, *J. Chem. Phys.* **82** 2136, (1985).

77. G.W.Farnell and E.L.Adler, in: *Physical Acoustics, Principles and Methods* edited by W.P.Mason and R.N.Thurston, Academic, New York, 1972, Vol. IX p.35

78. H.Ibach and D.L.Mills, Electron Energy Loss Spectroscopy and Surface Vibrations (Academic Press, New York, 1982).

79. J.P.Toennies, *J. Vac. Sci. Technol.* **A2**, 1055, (1984).

80. J.H.Wandass and J.A.Gardella,Jr., *Langmuir* **3**, 183, (1987).

81. S.R.Cohen, R.Naaman and J.Sagiv, *Phys. Rev. Lett.* **58**, 1208, (1987).

82. V.Vogel and Ch.Wöll, *Thin Solid Films* **159**, 429, (1988).

83. R.Zanoni, C.Naselli, J.Bell, G.I.Stegmann and C.T.Seaton, *Phys. Rev. Lett.* **57**, 2838, (1986).

84. J.R.Sandercock, in: *Light Scattering in Solids* edited by M.Cardona and G.Güntherodt, Springer-Verlag, Berlin, 1982, Vol. III, p.173.

85. X.D.Zhu, H.Suhr and Y.R.Shen, *Phys. Rev. B* **35**, 3047, (1987).

86. P.Guyot-Sionnest, J.H.Hunt and Y.R.Shen, *Phys. Rev. Lett.* **59**, 1597, (1987).

Vapor Deposition of Polyimide and Polyimide Precursors on Copper

R.N. Lamb[1], M. Grunze[2], J. Baxter[3], C.W. Kong[3], and W.N. Unertl[3]

[1]Division of Science and Technology, Griffith University,
  Brisbane 4111, Queensland, Australia
[2]Institut für Angewandte Physikalische Chemie, Im Neuenheimer Feld 253,
  Universität Heidelberg, D-6900 Heidelberg, Fed. Rep. of Germany
[3]Laboratory for Surface Science and Technology and,
  Department of Physics and Astronomy, University of Maine,
  Orono, ME 04469, USA

1. Abstract

Adhesion between metallic substrates and polymers is an important issue in microelectronic device fabrication and packaging. One of the most widely used classes of polymers are the polyimides which are employed as dielectric spacers in multichip modules and as α-particle barriers in charge sensitive memory devices. In order to study the polymer/metal interface with surface sensitive techniques, the polymer films have to have thicknesses on the order of a few nanometers to obtain chemical and physical information about the interface.

In this article we describe experiments in which ultra-thin films of polyimide were formed on polycrystalline and (111) single crystal copper substrates following heating of a vapor deposited layer of polyamic acid produced by codeposition of 4,4'-diaminodiphenyl ether (ODA) and 1,2,4,5- benzenetetracarboxylic dianhydride (PMDA). The adsorption of the pure components and their subsequent polymerization was followed, in situ, with x-ray photoelectron spectroscopy. Molecular adsorption of the monomers was studied to determine their thermal stability. The composition and adhesion of the thin polyimide film is influenced by fragmentation of precursors at the interface. In contrast to studies on silver and gold surfaces the polyamic acid interacts strongly with the copper substrates leading to a fragmented polyimide layer at the interface after imidization.

2. Introduction

In microelectronics, the application of temperature resistant polymers as insulating interlevel dielectrics has led to increasing interest in their preparation and subsequent adhesion to metal substrates [1]. The most popular of these dielectrics have been the polyimides and, in particular, those formed by the reaction of 4,4'-diaminodiphenyl ether (oxydianiline (ODA)) and 1,2,4,5- benzenetetracarboxylic anhydride (pyromellitic dianhydride (PMDA)). The chemical reaction leading to this polyimide is described in Figure 1.

The two main methods of film preparation are spin coating (SC) and vapor deposition polymerization (VDP). They differ in the way in which the film precursor is applied to the substrate. VDP is a solventless technique in which the evaporated monomers (ODA, PMDA) are codeposited directly onto the substrate whereas spin coating requires that the polymer precursor polyamic acid (PAA) is applied in a solvent.

Springer Series in Surface Sciences, Vol. 17
Adhesion and Friction   Editors: M. Grunze and H.J. Kreuzer
© Springer-Verlag Berlin, Heidelberg 1989

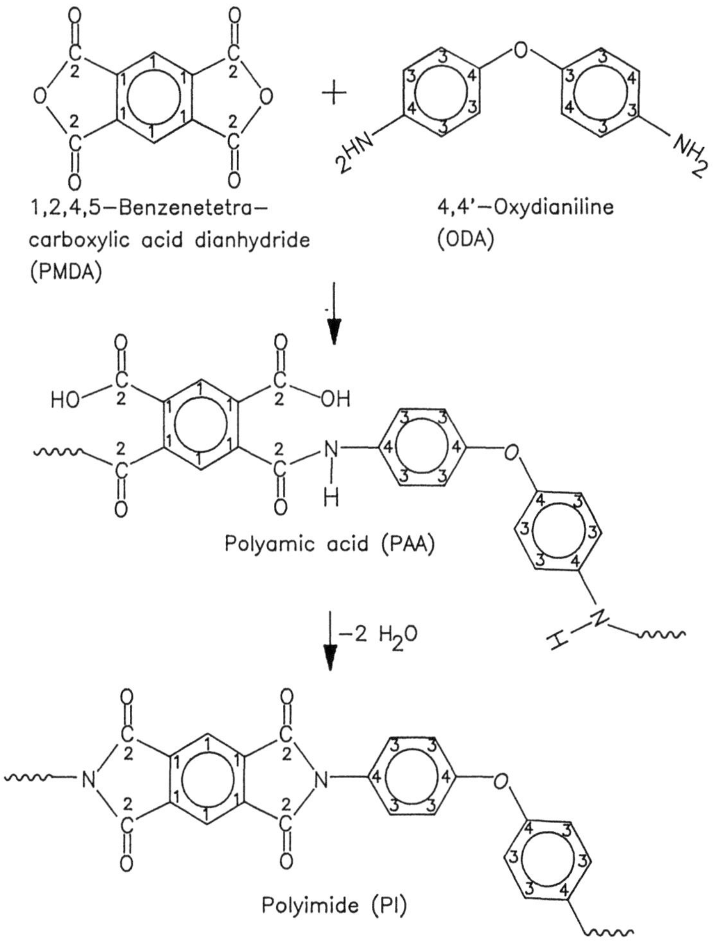

Fig. 1   Schematic representation of the reaction between PMDA and ODA to form polyimide. The numbers in the structural formulae are given to facilitate the interpretation of the x-ray photoemission data.

To evaluate the microscopic mechanism in polymer/substrate interfaces the undisturbed interface has to be probed with sufficient sensitivity. In the case of studies using electron spectroscopy, this requires that either the metal or polymer be sufficiently thin to allow elastic electron transport away from the interface and out to the vacuum for analysis. The relative ease with which this has been achieved in studies using metallized polyimides has led to interfaces for metal films deposited on PI being the main source of chemical interface information [2-8]. The difficulties associated with producing an (insulating) organic overlayer of sufficiently thin film thickness (usually d<10 nm) have only recently been overcome. Ultra thin SC films have been prepared on smooth, gold coated Si (100) wafers and the integrity of the film was investigated [9]. Alternatively VDP methods have been shown to routinely produce ultra thin and thin PI films on a variety of substrates [10-13].

It is this VDP class of interface which is of interest in this report. VDP for the formation of polyimide was first described in the production of thick (d > 1μm) films [15].   The ability to deposit ultra-thin (monolayer and submonolayer) and thin (monolayer < d <10 nm) films of <u>either monomer</u> on a clean substrate, held at any temperature and investigate the interfacial interaction, is unique to this method. In comparison, SC requires that the interface be heated to at least 373 K to remove the solvent prior to any spectroscopic investigation. This heating may intrinsically change the interface. Information available from vapor deposited monomer spectra also provide important reference data for the analysis of the more complex spectra arising from thin polymeric films.

The interfacial chemistry and adhesion is directly influenced by the way in which the interface is formed. Therefore, variations in the preparation of the polymer and/or metal will lead to polyimide/metal interfaces with different physicochemical properties. One particularly relevant example of this is the comparison between PI/copper interfaces formed by (i) copper deposition on cured polyimide (PI) and (ii) spin coating of the polymer precursor (PAA) onto a copper film prior to curing to form PI.

Kim and coworkers /2/ measured the adhesion strength by 90° peel tests for (i) and (ii).   They found, that in case (ii) adhesion is significantly enhanced as compared to copper deposited onto cured polyimide (i) and they attributed this to the difference in interfacial chemistry, i.e. chemical reaction between polyamic acid and bulk copper (ii) as compared to copper atoms or clusters interacting with cured polyimide. The difference in interface chemistry was also evident in cross-sectional TEM observations /2/. In the case of a sputter deposited copper film on a cured polyimide film, a sharp boundary was observed, whereas in the case of a polyimide/copper interface prepared by spin coating polyamic acid and subsequent imidization, cuprous oxide (Cu$_2$O) particles were found distributed in the polymer matrix beyond a precipitation free zone of ~8-20 nm thickness. That copper oxide particles are distributed over a thickness of ~500 nm was also recently found in an XPS study by Burrell et al. /14/ for spun-on polyimide films on copper substrates. In their XPS and IR - Reflection Absorption measurements they attributed a degradation or chemical modification of the thick polyimide films to copper oxide particle formation /14/.

Contrary to the observations made by Kim et al. /2/ on spun-on polyimide films, Kowalczyk et al. /3/ found no copper oxide particles in the polymer matrix in polyimide films produced by vapor deposition (VD). However, if prior to imidization a drop of the solvent N-methylpyrrolidone was applied to the vapor deposited polyamic acid film, copper oxide particle formation in the polymer matrix was observed. Their result clearly showed that the solvent provides mobility for copper oxide particles to diffuse into the polymer matrix, and suggest that initially the polyamic acid functionalities react with the substrate surface most likely forming carboxylate species which decompose during curing and act as a source for copper oxide formation. In this respect we recall our previous results on vapor deposited polyimide films on copper /13/ which showed that it is not possible to produce polyimide films of a thickness less than 4 nm. The thinner films (d<4nm) showed a strong deviation from the expected polyimide stoichiometry and exhibited x-ray photoelectron spectra characteristic for decomposition products.

The organization of this paper follows the reaction pathway for the formation of PI as described in Figure 1. Comparisons of monomer (ODA and PMDA) adsorption on single and polycrystalline copper substrates will be made and their effect on the formation of the polymer film assessed. The thermal stability of each adsorbate is examined in the context of the importance of the resulting polyimide as a temperature resistant polymer.

<u>3. Experimental</u>

The vapor deposition and XPS experiments were carried out with an apparatus described previously [11]. Tungsten wire was wrapped around quartz tubes containing the pure PMDA and ODA (Aldrich Gold Label) and heated resistively. At sublimation temperatures of between 373 K and 423 K, the resulting evaporation pressures ranged from 2×10^{-6} to 8×10^{-6} mbar. The "measured" deposition temperature was, however, sensitive to the positioning of the chromel alumel thermocouple in this design of oven.

The Cu substrate remained at room temperature during the codeposition process or was cooled down to 200 K in some instances for single monomer depositions. An arbitrary exposure scale, L=(background pressure in 10^{-6} mbar) x (exposure time in seconds) is used to indicate the extent of deposition. The polymer was formed by heating a codeposited layer of PMDA and ODA to temperatures ranging from 400 K to greater than 600 K for at least 1 hour. No effort was made to maintain a stoichiometric mixture of vapor fluxes during the experiment.

Sample cleaning prior to film deposition was carried out by heating the sample to 800 K in 13 mbar of O_2 followed by 3 mbar of H_2 to remove the surface oxygen. In some instances argon ion sputtering was also used. The Mg K_α x-ray source was operated at 100 W and an experimental resolution of 0.92eV was measured for Ag 3d emission. The electron binding energies (E_B) were calibrated against the Au $4f_{7/2}$ emission at E_B=84 eV.

Primary analysis of the XPS spectra arising from the organic depositions involves the determination of peak kinetic energy and integrated peak areas obtained after subtracting a linear secondary electron background. The mean stoichiometry of the film is calculated as a ratio of integrated areas by

$$\frac{N_1}{N_2} = \frac{I_1}{I_2} \; \frac{\sigma_2}{\sigma_1} \; \left(\frac{E_2}{E_1}\right)^{m \, - \, .73} \tag{1}$$

where N is the number of atoms/unit area for species 1 and 2 and I their core level intensities (proportional to peak area). The relative photoelectron excitation cross sections, σ, are 100, 285 and 177 for C 1s, O 1s and N 1s, respectively [16]. The transmission of the electron spectrometer ($E^{-0.73}$) and electron mean free pathlengths (λ) for organic materials are functions of the kinetic energy E. Combining these leads to the form for E shown in equation 1. The value for m (between 0.5 and 0.71) refers to the thick films ($d > 6 \, \lambda$) limit and leads to uncertainties of approximately 15% in the relative compositions [17]. Note also that final state effects such as shake-ups effectively borrow intensity from the primary peaks. Calibration spectra of pure monomers were used to assign additional features to incorporate them into the overall calculated stoichiometry.

The absolute peak binding energies are affected by charging within the film. The general trend was shifting to higher E_B with increasing film thickness. This can be explained by charging effects and/or a decrease in the final state screening of the photoionized molecules by metal electrons in the thicker films. Because of the indeterminate nature of these shifts, no corrections have been made to the data reported.

The distributions of bonding environments of the atoms within the films are contained within the XPS line shape. The complexity involved in analysis of lineshapes for polyimide films has been discussed recently [17]. In this paper, assignments of the components within the measured lineshapes are therefore made with reference to previous studies on PI and associated model compounds [11]. Calculated ratios of the various atomic and functional group species are compared with that expected from stoichiometry. The resulting deficits and/or additions are used to infer changes in the film composition.

Film thickness (d) was calculated by attenuation of the Cu 2p3/2 or Cu 3p3/2 intensity as

$$d = -\lambda \ln (I/I_0) \tag{2}$$

where λ is the mean free path length of the Cu 2p or 3p photoelectrons in the overlayer and has been assumed to have values of $\lambda_{2p}=0.8$nm ($E_K=320.2$ eV) or $\lambda_{3p}=1.8$ nm($E_K=1177.4$ eV) [18]. I_0 is the intensity measured on the clean surface. The values of d generated from (2) are considered reasonable estimates of true film thickness. A discussion of assumptions involved in (2) has been reported elsewhere [17]. In some instances, up to 99.99% attenuation of the Cu 2p signal (d< 7 nm) could be measured.

4. Results and Discussion

PMDA adsorption on Cu(111) at 200 K and subsequent thermal treatments.

In Figs. 2 and 3 we show the C 1s and O 1s spectra of a multilayer film of PMDA(a) and the spectral changes occurring when the film is heated stepwise to 573 K.

The C 1s photoemission of the condensed film (d > 7nm) exhibits two peaks, due to the phenyl ring carbon atoms (C1 in Fig. 1) at $E_B\sim287.1$ eV and carbonyl carbon atoms (C2, Fig. 1) at $E_B\sim291$ eV. These peaks are shifted by 1 eV to higher binding energies as compared to a monolayer of PMDA adsorbed on Cu(111) at 200 K indicating decreased final state screening of the core hole and/or charging of the film. The high binding energy tail on the carbonyl C 1s emission (Fig. 2a) is due to final state effects ("shake up") in the photoemission process. The exact origin of this shake up feature, i.e. its association with either the phenyl carbon or carbonyl carbon or both emissions, is not known. For the stoichiometry evaluation, the intensity in the high energy tail was added to the carbonyl C 1s intensity which, if this assignment is not strictly correct, leads to an overestimate of carbonyl groups in the PMDA film.

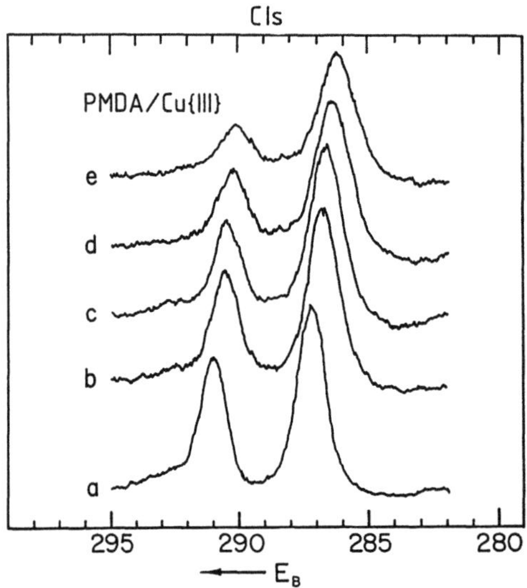

Fig. 2. C1s spectra for thermal treatment of a thick PMDA film deposited on Cu (111): (a) d > 7nm at 200 K; (b) d > 7nm following heating to 298 K; (c) d > 7nm following heating to 373 K; (d) d ~ 3.4nm following heating to 473 K; (e) d ~ 2nm following heating to 573 K.

The corresponding O 1s spectra for the condensed PMDA multilayer is shown in Fig. 3. The unresolved doublet arises from emission from the carbonyl oxygen atoms (O2) at E_B~534 eV and from the anhydride oxygen (O1) around E_B~535 eV.

The thick film of PMDA condensed at 200 K was subsequently heated to investigate the film stability. The resulting C 1s and O 1s spectra are shown in Figs. 2 b-e and 3 b-e. The reappearance of the Cu substrate signals following treatment at temperatures exceeding 373 K indicates either a considerable reduction in film thickness or film break up into PMDA aggregates. Accompanying this

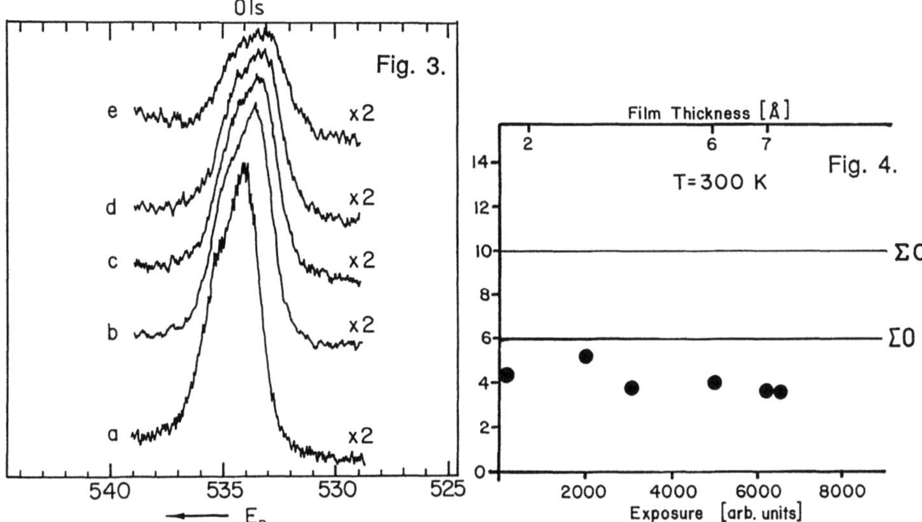

Fig. 3. O1s spectra for thermal treatment of a thick PMDA film deposited on Cu (111): (a) d > 7nm at 200 K; (b) d > 7nm following heating to 298 K; (c) d > 7nm following heating to 373 K; (d) d ~ 3.4nm following heating to 473 K; (e) d ~ 2nm following heating to 573 K.

Fig. 4.   Total $\Sigma C : \Sigma O : \Sigma C2$ ratio normalized to 10 PMDA carbon atoms for PMDA deposited on Cu(111) at 298 K

reappearance of Cu is peak broadening of the C 1s and O 1s emission, a shift to lower binding energy by (0.9 eV) and a disproportional reduction of the carbonyl C 1s and O 1s emission . In Fig. 4 we plot the integrated intensity of carbonyl carbon $\Sigma C_{C=O}$ (x) and total oxygen ΣO (o) normalized to total carbon ΣC as a function of temperature and mean film thickness estimated from the increasing intensity of the copper substrate emission. The results for oxygen are calculated in the "thick film" (i.e. m = 0.7) and "thin film" (i.e. m=0) limits. The lower stochiometry values indicated by the dots at the lower end of the vertical lines are obtained in the thick film limit, the thin film limit yields the upper dots. To the extent that the PMDA film remains uniform and homogeneous, the composition will be between these limits for d< 4 nm. For thicker layers, the thick films limit applies, the thin film limit is valid only for monolayer coverages (d«λ) .

The deviation from the expected composition of the condensed film ($\Sigma O=6$) reflects the uncertainties in our stoichiometry evaluation procedure, i.e. the definition of the correct background to be subtracted from the emission envelope and the errors associated with the value of m in eq. 1 used to correct for the energy dependence of the electron mean free path.

The change in lineshape occurs simultaneously with a loss of both carbonyl carbon (C2) and oxygen. The comparatively small decrease in carbonyl carbon compared to the oxygen loss suggests the latter may originate primarily from the anhydride (O1) functional group. Further temperature increases (T > 500 K) result in decreases of both carbonyl carbon and oxygen and signify increasing decomposition within the film. There is also evidence for partly ionic oxygen indicative of copper oxide formation at the highest temperature where a low binding energy shoulder appears in the 0 1s spectra at $E_B \sim$ 530 eV.

Thermal decomposition of the PMDA film is also evident in infrared reflection absorption (IRAS) measurements. A full account of the technique and the experiments is given elsewhere /19,20/. We give only a brief summary of the results of these IRAS studies. IR spectra of PMDA taken after adsorption at 223 K are representative of molecular PMDA. Heating the adsorbate layer to 300 K results in spectral changes indicating partial fragmentation at the interface, to which the IR reflection absorption technique is most sensitive. At this temperature, however, both carbonyl and anhydride stretching bands are still present in the spectra in addition to a new band at 1708 cm^{-1} which is indicative for a monodentate type bond of the PMDA fragment to the surface. Further heating to 373 K leads to the disappearance of the molecular vibration of PMDA. Considering the selection rules for IR-reflection absorption on metallic surfaces this could either be due to a reorientation of the molecule to lie flat on the surface so that the IR-active carbonyl and anhydride stretches are very weak due to image dipole cancellation or, in view of our XPS data, due to a decomposition of the molecule.

Decomposition of the PMDA molecule when adsorbed at room temperature was described in our previous work /13/. These results are summarized in Fig. 5 where we show the integrated intensity of oxygen (ΣO) normalized to the integrated carbon intensity (ΣC) as a function of exposure.

Comparison of room temperature adsorption of PMDA on copper and silver /11/ demonstrates that PMDA does not stick to highly fragmented PMDA. On copper, even at exposures as high as 6000 L there is little increase in the film thickness and the non stoichiometric $\Sigma C : \Sigma O$ ratio remains. Molecular adsorption does not therefore appear to occur in room temperature depositions even for very large exposures. Decomposition of PMDA, although to a lesser extent than on copper, has been reported for adsorption of PMDA on silver substrates at room temperature [11]. The partial dissociation involving loss of one CO molecule on silver did not inhibit multilayer molecular adsorption with further exposures /11/.

The thermal instability of the interface is highlighted by the obvious difference between PMDA adsorption on a Cu(111) substrate at room temperature where fragmentation occurs without multilayer formation and at 200 K where fragmentation does not occur (these data are not shown here) . It seems likely that an ultra thin stoichiometric PMDA film formed on a 200 K substrate would show similar dissociative behavior when heated to 298 K as does the PMDA adsorbed directly on a substrate held at this temperature. Clearly with increasing temperature the interface breaks down. As noted above, the degree of fragmentation at the interface is difficult to ascertain from the XPS measurements because of the thickness of the film and the unknown homogeneity of the layer. In our experiments we probe a weighted average composition of the film, i.e. we cannot distinguish between the signal arising from fragmented monolayers or thicker clusters or islands of PMDA. Heating the film might result in a breakup of the film and nucleation of PMDA crystallites. The observed decomposition products in our XPS data at estimated film thicknesses of d~3.5 nm therefore might arise from an inhomogeneous surface phase.

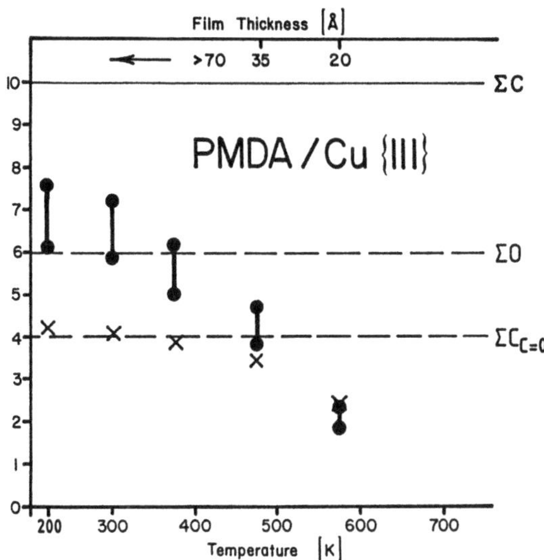

Fig. 5. Total $\Sigma C : \Sigma O : \Sigma C_{C=O}$ ratio normalized to 10 PMDA carbon atoms for thermal treatment of a thick film of PMDA deposited on Cu (111). For the O1s integrated intensities, the higher stoichiometric values were evaluated for the thick film limit, the lower values for the thin film limit (eq. 1). ΣO:•, $\Sigma C_{C=O}$: x

ODA adsorption on Cu(111) held at 200 K

The C 1s and O 1s spectra obtained following increasing exposure of ODA to a Cu(111) substrate at 200 K are shown in Figures 6-8. In the C 1s spectrum 6a the unresolved doublet originates from the C4 atoms (at higher E_B) and the lower binding energy shoulder corresponds to the C3 atoms. The binding energies remained constant up to exposures of ~120 L suggesting that charging is not important up to thicknesses of about ~ 4 nm. Shifts of 0.8 eV to higher binding energies occur for the thicker films from which spectrum (a) was taken. The lineshape remained constant above d>1.1 nm and indicates that ODA is adsorbing as a molecular species. The formation of multilayers of ODA without dissociation is also evident in the O 1s spectrum (Fig. 7a). The shift to higher E_B found for the C 1s emission at higher exposure was also apparent in both O 1s and N 1s spectra and reflects charging in the overlayer. The integrated O 1s and N 1s band intensities, normalized to the 12 carbon atoms of ODA, confirm a molecular stoichiometry for all film thicknesses.

Results for room temperature adsorption of ODA up to 1140 L on Cu {111} have been reported previously /13/. The spectra and the stoichiometry of the adsorbate phase suggested a complete dissociation of the ODA molecule upon adsorption.

Heating the thick ODA overlayer condensed at 200 K leads to sublimation and decomposition of the molecules. The C 1s, O 1s and N 1s spectra are shown in Figs. 6, 7 and 8b-e respectively. As the film thins the binding energies shift to lower values due to the elimination of charging. Spectrum d, recorded at room temperature, corresponds to a film thickness d~4 nm as judged by the reappearing intensity of the copper substrate photoelectrons.

As shown in the integrated intensities (Fig. 9), the relative concentration of nitrogen decreases at temperatures exceeding 330 K. Yet the O 1s concentration remains invariant, indicating a change in chemical composition of the film. This

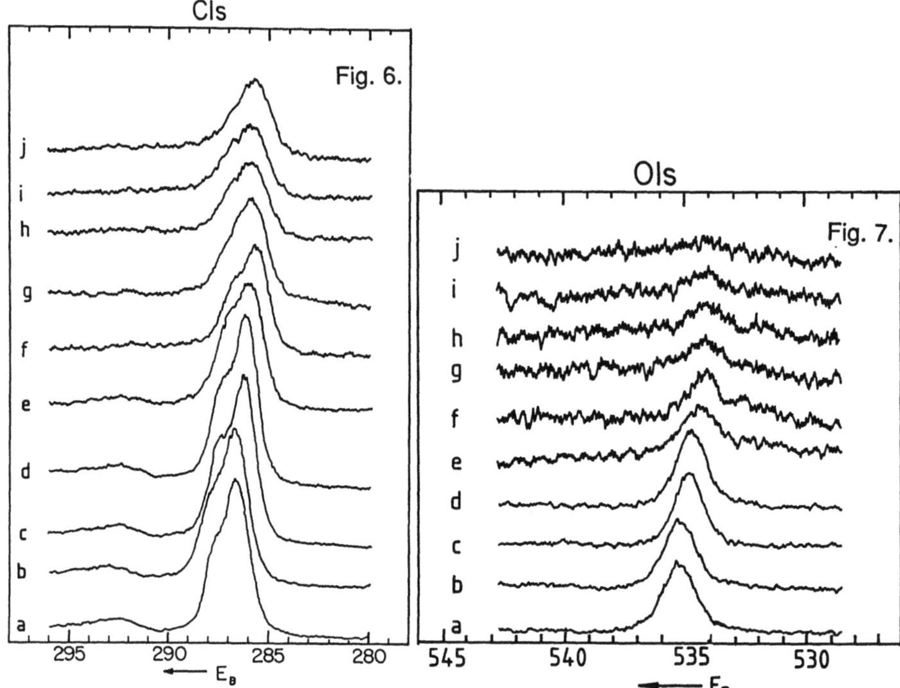

Fig. 6. C1s spectra for thermal treatment of a thick ODA film deposited on Cu(111): (a) Thick film d > 4 nm at 200 K; (b) d > 4 nm following heating to 223 K; (c) d > 4 nm following heating to 273 K; (d) d ~ 4 nm following heating to 298 K; (e) d ~ 2.5 nm following heating to 323 K; (f) d ~ 1.8 nm follow heating to 373 K; (g) d ~ 1.5 nm following heating to 423 K; (h) d ~ 1.3 nm following heating to 523 K; (i) d ~ 1 nmfollowing heating to 623 K; (j) d ~ 0.9 nm following heating to 723 K

Fig. 7. O1s spectrum of vapor deposited ODA on Cu(111) at 200 K and subsequent thermal treatment of a thick ODA film: (a) Thick film, d > 4 nm, at 200 K; (b) d > 4 nm following heating to 223 K; (c) d > 4 nm following heating to 273 K; (d) d > 4 nm following heating to 298 K; (e) d ~ 4 nm following heating to 323 K; (f) d ~ 4 nm following heating to 373 K; (g) d~1.5 nm following heating to 423 K; (h) d ~ 1.3 nm following heating to 523 K; (i) d ~ 1nm following heating to 623 K; (j) d ~ 0.9 nm following heating to 723 K

change is also evident in the line shapes of the C 1s, O 1s and N 1s spectra. The N 1s show only a decrease in intensity, but no clear shift in E_B, whereas a shift and a change in peak shape are observed for the C 1s data. This is paralleled by the shift of the O 1s band to 0.8 eV lower binding energy. Note, however, that the value for the O1s binding energy (533.5 eV) indicates a non-ionic form of oxygen and is not consistent with copper oxide formation at the ODA/metal interface, since this would give rise to a peak at E_B~530 eV /22/. The variation in stoichiometry of the ODA deposit occurs over the temperature range of 330 K - 420 K, at which temperature an average thickness of d~1.4 nm is reached. At T > 520 K further dissociation is indicated by the parallel decrease in the O 1s and N 1s intensities and results in an almost pure carbonaceous overlayer at T~700 K.

Adsorption of ODA at room temperature onto a polycrystalline copper foil was described previously /13/. Formation of a carbonaceous surface species was evident at the smallest exposure. It is only at the highest exposures (≥240 L) that

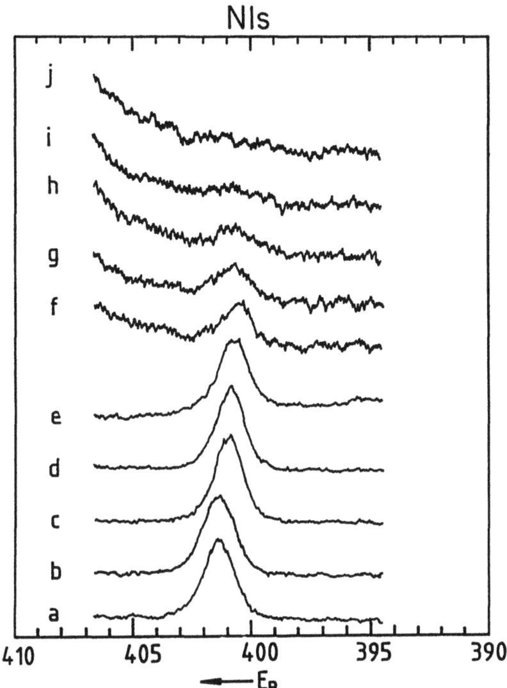

Fig. 8. N1s spectra of vapor deposited ODA on Cu(111) at 200 K and subsequent thermal treatment of a thick ODA film: (a) Thick film, d > 4 nm at 200 K; (b) d > 4 nm following heating to 223 K; (c) d > 4 nm following heating to 273 K; (d) d > 4 nm following heating to 298 K; (e) d ~ 2.5 nm following heating to 323 K; (f) d ~ 1.8 nm following heating to 373 K; (g) d ~ 1.5 nm following heating to 423 K; (h) d ~ 1.3 nm following heating to 523 K; (i) d ~ 1 nm following heating to 623 K; (j) d ~ 0.9 nm following heating to 723 K

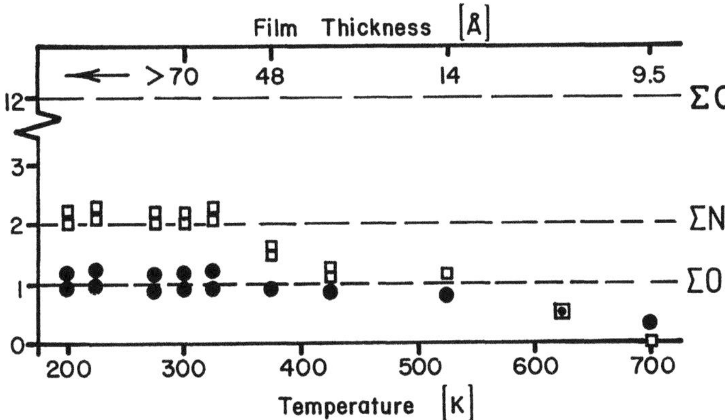

Fig. 9. Total C : O : N ratio normalized to 12 ODA carbon atoms for thermal treatment of a thick ODA film deposited on Cu(111): (l) ΣO: ΣN: $\square$; Upper and lower stoichiometry values correspond to the thin film and thick film limits of eq. 1, respectively.

the oxygen and nitrogen become discernible. The O1s spectra indicated, by a weak emission around $E_B\sim530$ eV, that, besides "organic" oxygen, an oxidic species is formed under these conditions. An evaluation of the stoichiometry of the overlayer after 1140 L exposure results in $\Sigma C{:}\Sigma O{:}\Sigma N$ of 12:0.9:0.8. If the oxidic oxygen at $E_B\sim530$ eV is excluded the ratio is 12:0.7:0.8. Clearly the layer did not consist only of molecular ODA. The fact that the carbon to oxygen ratio is close to the one expected (in particular where the oxidic species is included) but that nitrogen is lost in the initial adsorption process, suggested that the molecule dissociates and releases nitrogen or nitrogen carbon entities. Aniline is one possibility which would also leave an oxianiline species remaining at the surface. Such a reaction scheme has been postulated for the interaction of ODA with silver surfaces [11] but is speculative and needs to be corroborated by other experimental techniques. Since the film thickness after room temperature adsorption does never exceed 0.7 nm it is clear that the dissociation reaction occurs at the interface and precludes multilayer adsorption.

Infrared reflection absorption spectroscopy data for the thin ODA films on clean polycrystalline copper [20] reveal the complete loss of characteristic molecular ODA vibrations at $T\sim350$ K, suggesting that ODA is either completely desorbed or converted into an amorphous carbonaceous overlayer. The disappearance of molecular vibrations in the infrared absorption spectra however, does not necessarily mean that the surface layer loses oxygen and nitrogen completely, as also indicated by the persistence of the N 1s and O 1s photoemission signal up to $T\sim700$ K. It is therefore probable that the nitrogen and oxygen fragments are incorporated into the organic overlayer but are not active in the infrared absorption experiments. The orientation of the dynamic dipole moments of the fragments with respect to the metal image plane determines their absorbance. The XPS binding energies of the N 1s and O 1s emission indicated that the nitrogen and oxygen atoms in the fragments are not interacting directly with the metal substrate, since this would lead to a further shift to lower binding energy values.

<u>Summary of monomer adsorption on single crystal and polycrystalline Cu</u>

The PMDA adsorption experiments were only carried out on a Cu(111) surface. The low sticking coefficient at room temperature was accompanied by dissociation of the adsorbate molecules with the loss of carbonyl groups. Multilayer molecular adsorption occurs for deposition on Cu(111) substrates cooled to 200 K. Heat treating a thick film produced sublimation and decomposition within the film.

Molecular adsorption of ODA occurs on Cu(111) surfaces cooled to 200 K. Heating such a layer leads to sublimation and decomposition resulting in the loss of nitrogen and nitrogen carbon entities. Dissociative adsorption occurs at room temperature on both Cu(111) and polycrystalline Cu substrates.

The copper surface appears to destabilize the adsorbates to a greater extent than does polycrystalline silver or gold /11,21/. The inability to form multilayer PMDA or ODA on copper at room temperature highlights this. Obviously copper has a much stronger affinity to fragment the polyimide constituents than the two other noble metals studied in our laboratory. However, from our XPS results we cannot conclude that the copper substrate influences the chemical composition of the films when their thickness is several monolayers. Thickness estimates are based on the assumption that the film is <u>homogeneous</u> with uniform thickness. On silver substrates thermal desorption and IR-Reflection Absorption measurements /20/ showed conclusively that thick ODA and PMDA layers condensed at low temperatures form crystallites when the thick films are heated, thus leading to a non-uniform thickness of the layer. In this case, XPS would simultaneously probe both the thick crystalline PMDA and ODA and the decomposed surface layer so that the spectra would reflect an average composition. This inhomogeneous composite layer could lead to the changes in average stoichiometry as well as to the broadening of the peaks observed in the heating experiments. Although this is the most likely interpretation of our results, the possibility that copper atoms or ions diffuse from the interface into the thick PMDA and ODA films and alter their chemical composition cannot be ruled out . Further experiments need to be carried out to resolve this question.

<u>Codeposition of PMDA and ODA on Cu(111)</u>

The C 1s, O 1s and N 1s spectra for PMDA and ODA codeposited on to Cu(111) at room temperature are reproduced in Figs. 10-12(a-d). This layer was then heated slowly in vacuum and resulted in a reduction in film thickness and changes within the XPS spectra as shown in Figs. 10-12(e-j). The variation in spectral lineshape associated with the imidization reaction and the method for deconvolution of these has been reported in detail elsewhere [17]. A similar analysis is used in the following.

Spectra 10-12(a) are taken from the "clean" Cu (111) surface which, however, shows a small oxygen contamination. Subsequently PMDA and ODA were

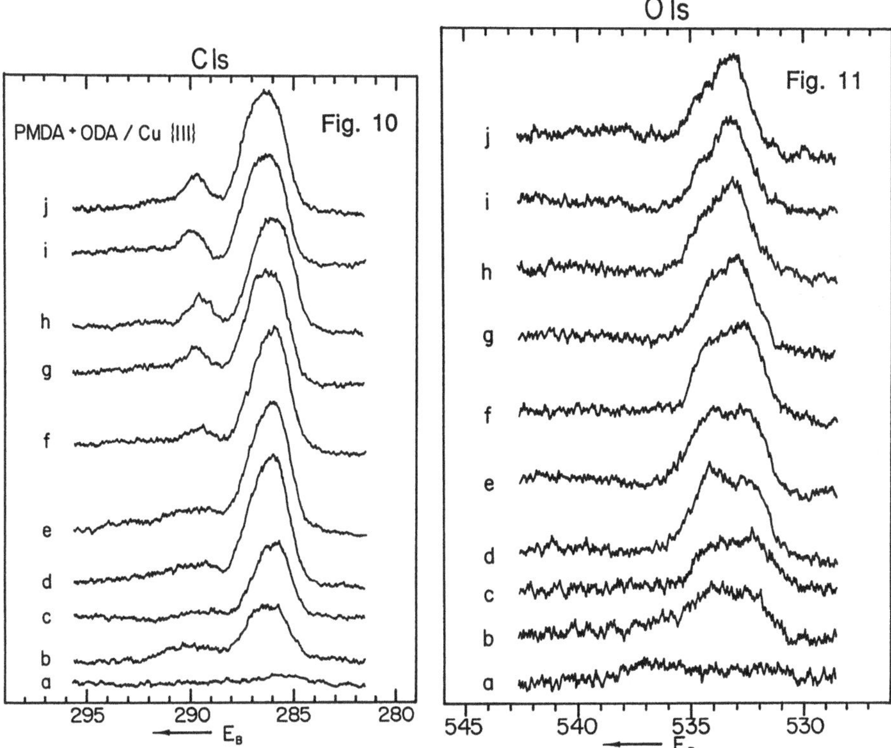

Fig. 10. C1s spectra for the coadsorption of PMDA and ODA on Cu(111) and subsequent heating to form polyimide: (a) Clean substrate (at 298 K); (b) L = 60 (at 298 K) d~1.6 nm; (c) L = 180 (at 298 K) d~1.7 nm; (d) L = 480 (at 298 K) d~7.3 nm; (e) Following heating to 329 K for 1hr; (f) Following heating to 373 K for 1hr; (g) Following heating to 423 K for 12hrs; (h) Following heating to 473 K for 1.5 hrs; (i) Following heating to 573 K; (j) Following heating to 673 K

Fig. 11. O1s spectra for the coadsorption of PMDA and ODA on Cu(111) and subsequent heating to form polyimide: (a) Clean substrate (at 298 K); (b) L = 60 (at 298 K) d~1.6 nm; (c) L = 180 (at 298 K) d~1.7 nm ; (d) L = 480 (at 298 K) d~7.3 nm; (e) Following heating to 329 K for 1hr; (f) Following heating to 373 K for 1hr; (g) Following heating to 423 K for 12hrs; (h) Following heating to 473 K for 1.5 hrs; (i) Following heating to 573 K; (j) Following heating to 673 K

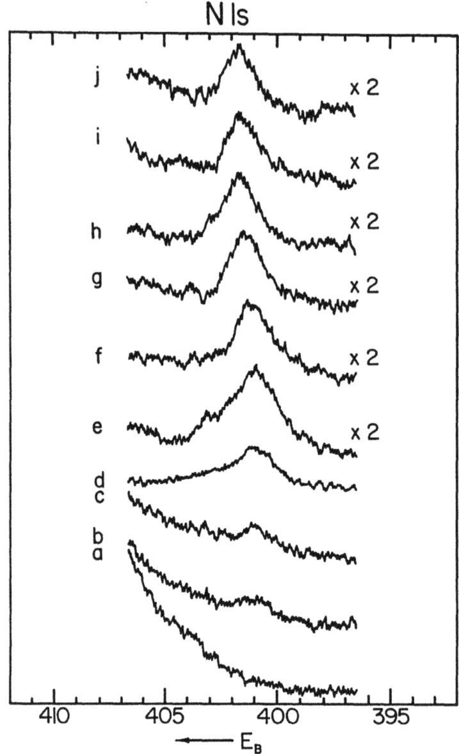

Fig. 12. N1s spectra for the coadsorption of PMDA and ODA on Cu(111) and subsequent heating to form polyimide: (a) Clean substrate (at 298 K); (b) L = 60 (at 298 K) d~1.6 nm; (c) L = 180 (at 298 K) d~1.7 nm; (d) L = 480 (at 298 K) d~7.3 nm; (e) Following heating to 329 K for 1hr; (f) Following heating to 373 K for 1hr; (g) Following heating to 423 K for 12hrs; (h) Following heating to 473 K for 1.5 hrs; (i) Following heating to 573 K; (j) Following heating to 673 K

codeposited onto the surface held at room temperature, i.e. both evaporators were operated simultaneously.   The first obvious difference from deposition of the pure monomers is that after relatively small exposures a multilayer is formed.   This result, that the condensation coefficient for codeposition is by orders of magnitude higher than for pure monomer adsorption, occurs on all substrates studied so far, i.e. Cu, Ag, Au , Ni and SiO_2. The high condensation coefficient remains for thick multilayer film formation; in the example shown here, a formal exposure of 480 L is sufficient to produce a film exceeding a thickness of d >7 nm.

The codeposited layer clearly is not a mixed phase of ODA and PMDA, but consists of polyamic acid (see Fig. 1) as has been discussed previously /11,15/. The C 1s, O 1s and N 1s spectra of the codeposited films (b,c,d) taken after sequential exposures have different lineshapes, indicating that the chemical composition varies and is more complex than simple polyamic acid. A stoichiometry analysis showed that spectra b and c have a pronounced deficit of nitrogen, i.e. for spectra c the ratio between $\Sigma C : \Sigma O : \Sigma N = 22 : 6.4 \pm 0.7 : 1.13 \pm 0.05$ compared to the stoichiometric ratio of 22:7:2 for polyamic acid. For spectra (d) an excess of ODA is indicated from the excess of nitrogen and slight deficit of oxygen ($\Sigma C : \Sigma O : \Sigma N = 22 : 5.5 \pm 0.6 : 2.9 \pm 0.25$).   Also consistent with an excess of ODA is the $\Sigma N : \Sigma C2 : \Sigma O$ ratio of 2:2.5:3.5 compared to 2:4:7 for polyamic acid [16]. The higher relative carbon content of ODA with respect to PMDA results in a low value for the oxygen stoichiometry.

The initial deficit of nitrogen is not consistent with a PMDA excess because this would also require excess of oxygen due to the higher oxygen/carbon ratio in PMDA as compared to ODA. The result, however, resembles our room temperature adsorption studies of ODA /13/, where a pronounced nitrogen deficit of the adsorbed phase was found. Since nitrogen is not contained in this adsorbate phase, some molecular nitrogen species must have been desorbed. The mechanism by which this occurs for pure ODA adsorption and ODA/PMDA codeposition is not known, but the similarity of these two cases highlight the fact that the interfacial chemistry is determined by the functional groups of the organic adsorbates.

The changing lineshapes of the C 1s, O 1s and N 1s spectra b-d following sequential deposition reflect the chemical changes in the ODA/PMDA codeposited layer. The carbon 1s spectrum (b) shows a broad phenyl carbon emission around $E_B = 286$ eV and a broad emission band around $E_B = 290$ eV indicative for carbonyl groups in a variety of chemical environments. In the O 1s emission the high binding energy shoulder at 534 eV is due to hydroxyl oxygen in the polyamic acid deposit. However, only at the higher exposures of spectra (d) does the C 1s, O 1s and N 1s emission became similar to those of solventless polyamic acid produced on silver and gold surfaces by codeposition of PMDA and ODA. The carbonyl intensity is substantially suppressed and broadened while the hydroxyl O 1s emission becomes more pronounced. In the N 1s spectrum (d) the increased width of the peak with the higher binding energy tail indicates that the ODA amino groups are reaction centers in the formation of the polymer (see Fig. 1).

It is the initial bonding of the PAA to the substrate that ultimately determines the fundamental adhesion of the polymer. In previous studies of VDP of polyimide on silver [11] it was postulated that the fragmented monomers acted as anchors for the resulting polymer chain. If this hypothesis is extended to the present case then at the interface only PMDA would contain any functional groups which could react with molecular ODA. As noted earlier, ODA completely dissociates at the room temperature Cu(111) interface with loss of functional groups. The 1.6 nm films of codeposited PMDA and ODA may therefore contain no interfacial ODA. From the lineshape of the C1s spectrum corresponding to 60 L exposure it is clear that the broad carbonyl peak is more pronounced compared to films of increasing thickness. This may originate from PMDA at the interface.

The fact that multilayers of codeposited PMDA and ODA form at room temperature while neither PMDA nor ODA alone form multilayers suggests that a chemical reaction occurs during codeposition. Accordingly the resulting XPS spectra are not simply a composite of the monomer spectra. We are unable to determine from the experiments reported here whether the PAA forms at the interface or in the vapor prior to adsorption. The relative size of the polymer does however suggest that Van der Waals forces will play some role in initial adhesion. This would of course increase the initial sticking coefficient.

As the film is heated, the hydroxyl O 1s at $E_B = 534$ eV decreases relative to the lower binding energy O 1s emission and the carbonyl C 1s emission is resolved above the background. Concomitant with this is a narrowing of the N1s emission and a shift to higher E_B indicating the onset of imidization. Heating at 573 K and above (Figures 10-12 i-j) does not further change the spectral lineshape which indicates that the curing to form polyimide is complete. The C 1s spectra do not, however, indicate pure polyimide. The peak centered around 286.2 eV does not show the characteristic splitting due to C3 (lower binding energy) and C1 and C4 at higher E_B. Similarly, in the O 1s there is an increase in the lower E_B peak relative to the higher binding energy shoulder. The N 1s peak exhibits a very weak shoulder at lower binding energy, but otherwise shows very little change.

The ratio of the integrated C:O:N intensities during heating is shown in Figure 13 a. As the film is heated there is definite loss of nitrogen due to desorption of excess unreacted ODA. Increasing temperatures lead to imidization and a

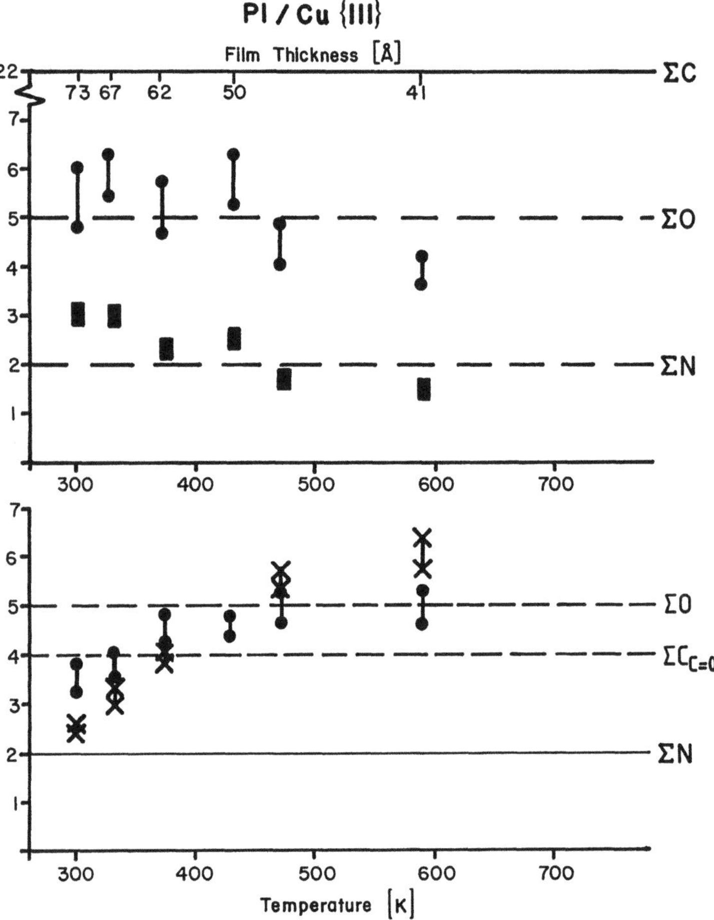

Fig. 13.(a) Total C:O:N ratio normalized to 22 carbon atoms for codeposited PMDA and ODA on Cu(111) at 298 K and subsequent heat treatment. (l) ΣO:• (n) ΣN: ■; (b) Total O:C2:N ratio normalized to 2 nitrogen atoms for codeposited PMDA and ODA on Cu(111) at 298 K and heat treated; ΣO:• ; ΣC_{C=O}: ✕

decrease in oxygen content. At temperatures between 373 K and 473 K there appears to be stoichiometric polyimide.

Figure 13 (b) summarizes the ratio of ΣN:ΣO:ΣC2 integrated intensities and serves to highlight some of the subtleties associated with the reaction. For T>373 K the total N:O ratio remains essentially stoichiometric (2:5). In comparison the carbonyl carbon C2 appears to increase as imidization progresses. Much of this initial increase can be attributed to the procedure for background subtraction. Because the broad carbonyl emission in PAA is difficult to separate from the shake-up transitions the latter were initially included in the overall C2 intensity. As the temperature increases, the imidization results in the reappearance of a well defined carbonyl band, which makes the integration much easier. Also, if the phenyl carbons have measurable intensity shake-up transitions, then the carbonyl content will be overestimated at all stages [16]. It is unlikely, however, that the observed large increase in carbonyl can be accounted for in this way. The increase in total carbon relative to nitrogen and oxygen is thus caused by an excess of carbonyl carbon C2.

The lineshape changes indicate that the film produced at 573 K is not pure polyimide but contains a mixture of PMDA and to a lesser extent ODA fragments (the shoulder in the N1s is an indication of this). These excess constituents could be in the form of trapped molecules or terminal groups, but branching and crosslinking might also explain the results. Similar results in very thick films on silver have been reported previously. The dissociation of single monomers on the substrate also suggests that the bulk of the additional carbon in these films could be due to carbonyl carbon arising from the PMDA like fragments in the interface.

Similar results have been reported on polycrystalline Cu[13], for a 4.2 nm thick film. This thicker overlayer did show a much more stoichiometric ratio of C:O:N. The lineshape, however, was also indicative of the presence of ODA and PMDA like fragments.

The differences in the polyimide films formed on polycrystalline and (111) copper substrates may then be attributed in part to the interface reaction i.e. to differences in the relative reactive sticking coefficients of PMDA and ODA. If the reactive sticking coefficients of PMDA and ODA leading to carbonyl and nitrogen loss, respectively, are similar on polycrystalline copper but ODA decomposition under the release of nitrogen containing fragments is favoured on Cu (111), the resulting polyimide/copper interface composition will reflect this through an excess of carbonyl groups. Despite having started in the thick polyamic acid film with an excess of ODA , curing has led to an excess of PMDA like species. This is consistent with PMDA and ODA adsorption at room temperature, where the latter is completely dissociated. On polycrystalline Cu the ODA and PMDA were present in much more comparable amounts and the resulting polyimide film composition reflected this .

We can only speculate about the bonding at the polyimide/copper interface at this time. Given the behaviour of PMDA and ODA at T~ 575K, we expect that there would in fact be a loss of carbonyl and a breakdown of the polyimide/copper interface into an amorphous carbon layer. However, the polymer has a stabilizing effect and the chemistry at the interface may also be altered due to this.

In previous work /13/ the subject of the minimum possible polyimide film thickness was addressed. It was reported that polyimide films thinner than ~4 nm were not possible on polycrystalline Cu while ultra thin films of about 1 nm were possible on polycrystalline Ag. Fragmentation at the interface could be a major cause of this difference. Clearly monomer dissociation is much more pronounced on copper. Figures 9-11a-d show that with increasing exposure an ODA excess is only realized at films with d>2 nm. If, as suggested previously, this is due to PMDA adsorption being a possible base for future polymer growth then the minimum thickness must take into account the excessive fragmentation at the interface. On silver there is much less fragmentation and the resulting fragments are all available for polymerisation.

In the context of understanding the transmission electron microscopy studies of spun on and vapor deposited polyimide films on copper /2,3/ we note that a strong initial reaction occurs between the substrate and both the monomers and polyamic acid. This initial reaction involves covalent and, as has been noted in the PMDA temperature profile, oxidic (partially ionic) bonded copper surface atoms. These then may provide a source for copper ion formation and, subsequent migration into the polymer matrix in the presence of the solvent.

5. Summary

It has been shown that thin polyimide films can be prepared by vapor deposition polymerisation on Cu substrates. The resulting film composition can be related to the initial monomer adsorption. Copper appears to destabilize the adsorbates at room temperature much more than that seen previously on Ag substrates. The minimum polyimide film thickness possible on Cu is found to be about 4 nm and can be explained by the increased degree of fragmentation at the interface.

Acknowledgments

This work was supported in part by the Office of Naval Research, and the National Science Foundation Grant No. DMR-8403831 and the Laboratory for Surface Science and Technology. R. L. thanks the Deutscher Akadademischer Austauschdienst for a stipend.

References

1. Polymer Materials for Electronic Applications, A.C.S. Symp. Ser., 184 (1982)
2. Y.-H. Kim, G.F. Walker, J. Kim and J. Park, J. Adhesion Sci. Tech. 1 331 (1987).
3. S.P. Kowalczyk, Y.H. Kim, G.F. Walker, J. Kim, Appl. Phys. Lett., 52, 375 (1988).
4. N.J. Chou and C.H. Tang, J. Vac. Sci. Technol. A, 2, 751 (1984).
5. J.L. Jordan, P.N. Sanda, J.F. Morar, C.A. Kovac, F.J. Himpsel and R.A. Pollak, J. Vac. Sci. Technol. A, 4, 1046 (1986)
6. P.N. Sanda, J.W. Bartha, J.G. Clabes, J.L. Jordan, C. Feger, B.D. Silverman and P.S. Ho, J. Vac. Sci. Technol. A, 4, 1035 (1986)
7. P.S. Ho, P.O. Hahn, J.W. Bartha, G.W. Rubloff, F.K. LeGoues and B.D. Silverman, J. Vac. Sci. Technol. A, 3, 739 (1985)
8. F.S. Ohuchi and S.O. Freilich, J. Vac. Sci. Technol A, 4, 1039, (1986)
9. J. Russat, Surf. Int. Anal., 11, 414 (1988)
10. M. Grunze and R.N. Lamb, Chem. Phys. Letts., 133, 283 (1987).
11. M. Grunze and R.N. Lamb , Surf. Sci., 204 (1988) 183.
12. M. Grunze and R.N. Lamb, J. Vac. Sci. Technol. A, 5, 1685 (1987)
13. M. Grunze, J.P. Baxter, C.W. Kong, R.N. Lamb, W.N. Unertl and C.R. Brundle, "Vapor Phase Growth of Polyimide Films" AVS National Symposium, Topical Conference "Deposition and Growth: Limits for Microelectronics", Anaheim, Nov. (1987).
14. M. C. Burrell, P. J. Codella, J. A. Fontana, J. J. Cheva, M. D. McConnell, J. Vac. Sci. Technol A, 7, 55 (1989).
15. J.R. Salem, R.O. Sequeda, J. Duran, W.Y. Lee and R.M. Yang, J. Vac. Sci. Technol. A, 4, 369 (1986).
16. J.H. Schofield, J. Electron. Spectros., 8, 129 (1976)
17. R.N. Lamb, J. Baxter, M. Grunze, C.W. Kong and W.N. Unertl, Langmuir, 4, 249 (1988)
18. D.T. Clark in "Chemistry and Physics of Solid Surfaces," Vol. 11, Ed. R. Vanselow, CRC Press, Boca Raton (1979)
19. M. Grunze, W. N. Unertl, S. Gnanarajan and J. French, Mat. Res. Soc. Symp. Proc. Vol. 108, (1988) 189.
20. S. Gnanarajan, Ph. D. Thesis, University of Maine (1989).
21. T. Strunskus, Diplomarbeit, Univ. Heidelberg (1988).
22. J. P.Baxter, M. Grunze, C.W. Kong, J.Vac. Sci. Technol. A6 (1988) 1123

Metallized Polymers – An Electron Induced Vibrational Spectroscopy Approach

J.J. Pireaux, M. Vermeersch, N. Degosserie, C. Grégoire, Y. Novis, M. Chtaïb, and R. Caudano

Facultés Universitaires Notre-Dame de la Paix,
Laboratoire Interdisciplinaire de Spectroscopie Electronique,
Rue de Bruxelles, 61, B-5000 Namur, Belgium

ABSTRACT In the search for better characterization and understanding of the interaction between organic and metallic films, this paper will review recent trials of application of a new spectroscopy - namely, HREELS (High Resolution Electron Energy Loss Spectroscopy). The incipient phases of the metallization of some selected polymers (polyimide, polyvinyl alcohol, polyacrylic acid and polystyrene) have been studied by monitoring the vibrational spectra of the polymers during well controlled aluminum evaporation from a Knudsen cell. In comparison with XPS (X-ray Photoelectron Spectroscopy), HREELS can detect hydrogen (via C-H or O-H stretch bands), deduce the site of the metallic atom deposition, and the molecular conformation of the metal-polymer complex at the interface. In the case of aluminum deposition on the chosen polymers, we show that different successive chemical reactions can be distinguished. When the polymer contains oxygen atoms, aluminum reacts readily to form a metal-oxygen-polymer complex, with an increasing reactivity going with C=O > C-OH. In all cases, the final surface composition (after a few Å of Al coverage, at room temperature) is dominated by Al-O, Al-C and O-Al-O species.

INTRODUCTION

Interface phenomena generally develop on a scale of a few angstroms, i.e. within a few atomic layers. The technological importance of interfaces - of any kind, between any two (or more) layers of metals, semiconductors, insulators, or polymers - requires from scientists the development and better understanding of characterization techniques capable of reaching the few angstrom detection level. Better electrical or mechanical (adhesion) properties of the interfaces will indeed progress as our knowledge of their geometrical, atomic and chemical compositions increases and our experience of tailoring at will those parameters. Transmission electron microscopy and (synchrotron) X-ray photoelectron spectroscopy (XPS) have so far contributed significantly to this goal. Still, another dimension to this analysis would be welcome.

High-resolution electron energy loss spectroscopy (HREELS) has been demonstrated to have such potentiality. For adsorbed molecules on metal surfaces, electron-induced vibrational spectra are sensitive to adsorption sites : information is available on the composition, the chemical reaction, and the structure or morphology of the interface [1].

Springer Series in Surface Sciences, Vol. 17
Adhesion and Friction   Editors: M. Grunze and H.J. Kreuzer
© Springer-Verlag Berlin, Heidelberg 1989

Metallized polymers are just one class of interfacial materials : for those systems, HREELS is capable of detecting hydrogen species (as CH_x ($1 \leq x \leq 3$) or OH entities), and even of easily differentiating between aromatic and aliphatic CH_x species. Should the electronic dipolar selection rule play a role in determining the HREELS vibrational intensities (as in IR spectroscopy), then it is possible to extract information from a layer thickness in the range of 100 Å [2]. However, HREELS is intrinsically very surface sensitive, as it uses as a probe a low energy electron beam ($E \simeq 5$ eV) [3].

In fact, the application of HREELS to characterize polymer surfaces is just emerging from its infancy [3]. Many fundamental questions need still to be solved in order to routinely apply HREELS quantitatively to polymer surfaces. But that did not restrain the research workers from going beyond those problems and tackling the difficult system that represents a metal-polymer interface. Data already published on palladium-, chromium-, and aluminum-polyimide interfaces [4, 5] were so promising that a systematic HREELS investigation of the incipient interface formation between an evaporated aluminum film and model polymers has been undertaken : it is the purpose of this paper to review those results on polyvinyl alcohol, polyacrylic acid [6], and on polystyrene [7]. Studies on polyethylene terephthalate are currently being conducted. Those polymeric materials have been chosen to present single chemical functionalities - as opposed to polyimide - which are respectively, an alcohol function, an acid entity, a phenyl pendant group, and other ether linkages.

Our review will begin with a short summary of the important contribution of XPS to the study of metal-polymer interfaces; we will indeed stress later on the complementarity between the XPS and HREELS techniques. A short presentation of the HREELS experiment itself will be made. Then, our choice of aluminum as the evaporated metal will be justified, before we present the HREELS results on the selected polymeric materials. Finally, we shall try to draw some general conclusions from the available XPS and HREELS measurements.

1. The X-Ray Photoelectron Spectroscopy Contribution

In this section, we have chosen to summarize some XPS contributions to the study of metal-polymer interfaces : this choice - which might appear arbitrary - is justified as follows:

(1) BURKSTRAND's work appears as the first systematic XPS study of the interface between different evaporated metals (copper, nickel and chromium) on various polymers (polystyrene, polyvinyl alcohol, polyethylene oxide, polyvinyl methyl ether, polyvinylacetate, and polymethylmethacrylate) [8].

(2) DeKOVEN and HAGANS' contribution is specifically related to the XPS study of thin films of Al on polyacrylic acid [9].

(3)   ATANASOSKA   et al.'s very recent paper reports on high resolution XPS analysis of the aluminum/polyimide interface formation [10].

XPS analysis and adhesion strength measurements for different metal/polymer interfaces have proved that : (a) polymers containing single bonded or multibonded oxygen atoms behave very differently ; (b) metallic atoms have different mobilities on the polymer surface ; (c) the metal is more tightly bonded to the polymer through a metal-oxygen carbon complex than if oxygen is not present, as on clean polystyrene ; (d) on PMMA, both single- and double-bonded oxygen atoms should interact with the adsorbed metallic atoms ; (e) the ratio of metallic atoms to oxygen ones on the polymer surface is about 1 : 2 [8].

The study of the aluminum-poly(acrylic acid) interface revealed that the COOH groups appear to be very reactive towards the metal atoms, with both oxygen and carbon atoms reacting with the deposited Al and with the formation of Al oxide or carbide species on the polymer surface. If before the metallization there was evidence of hydrogen bonding on the PAA and a larger than statistical number of carboxylic acid groups pointing away from the surface towards the bulk, the metal evaporation was probably inducing some polymer surface damage [9].

The Al-polyimide interface formation appears more complex : within the first Å of Al deposition, the bonding occurs at carbonyl sites, the carbon-oxygen double bond weakens, but preserves its character : in fact, some kind of resonance hybrid having Al ions in the vicinity of the $C = O$ bonds is suggested. With higher Al coverage, a C-O-Al complex compound is formed, as suggested by the observation of strong Al-O bonds. With further metal deposition, the carbonyl signal does not completely vanish, as 50 % of the initial intensity is still observed; metallic Al is detected, indicating the overlayer is highly inhomogeneous, with Al island development [10].

High resolution electron energy loss spectroscopy is expected to contribute to an even better understanding of the metal-polymer interaction. In comparison with XPS, the HREELS vibrational spectra detect hydrogen species easily (through e.g. the $C-H_x$ ($1 \leq x \leq 3$) or OH stretching bands); they distinguish aliphatic and aromatic CH_x species, and contain information on the stucture and the morphology of the polymer surface. It is then well suited to providing complementary information on the chemical and geometrical issues of an interface formation, when a metallic film is evaporated onto the polymer surface.

2. The HREELS Experiment

Located in a UHV chamber (base pressure $< 1 \times 10^{-10}$ Torr), the electron energy loss spectrometer consisted of two hemispherical electrostatic deflectors, used respectively as a monochromator of the incident beam, and an analyzer of the electrons backscattered from the sample surface. An electron impact energy (not corrected for work function difference) of 6.0 or 7.0 eV was used ; the spectrometer was configured in the specular geometry (impact angle = analysis angle), with a nominal resolution of about 10 meV.

Compared to clean or adsorbate-covered metal or semiconductor surfaces, polymers are rather complex materials, rarely chemically pure, rarely structurally well defined. A large number of different chemical groups in a polymer will be excited by the incident electron beam and show up as a forest of vibrational bands ; in order to resolve as many structures as possible, the highest instrumental resolution is required. Practically, however, the disorder and roughness of most of the polymer surfaces broaden the experimental results; a resolution - as measured by the full width at half maximum of the elastic peak - of about 10 meV ($\simeq$ 80 cm^{-1}) is currently obtained when another particular problem can be counterbalanced.

Organic materials are generally insulating; they offer to the experimentalist the build up of an electrostatic charge on their surfaces, when they are subjected to a charged particle beam : this charging effect often completely ruins the recording of spectra involving ions or electrons. If a modified "flood-gun" technique can be used to record HREELS spectra from polymer surfaces [11], it is better - in order to reduce radiation damage - to study thin (50 - 250 Å) polymeric films deposited onto a conducting substrate : spin coating, casting from a solution, or a Langmuir - Blodgett type of polymer film preparation will most of the time eliminate the charging problem.

3. The Polymers and the Metal Evaporation

The polymeric films and the evaporated metal were chosen to meet a technological opportunity, and the desire to understand the fundamentals of the metal-polymer interface formation.

Polyimide is probably one of the most often studied polymers nowadays; its actual and future high-technology applications, resulting from its particular dielectric, mechanical, and thermal properties, are not unrelated to this interest. However, polyimide is a rather complex material, whose chemistry and structural properties are not yet completely understood. In order to proceed more gradually, other much simpler polymers were also chosen : they were selected to contain separately simple and single chemical functions (Fig. 1) : polyvinyl alcohol, polyacrylic acid and polystyrene do indeed present respectively a single alcohol (-OH), acid (-COOH), or aromatic (-C$_6$H$_5$) function, whereas polyimide contains aromatic rings, ether linkages (C-O-C), carbonyl (C=O) groups, -C-N functions, fused rings, etc. In order to discover what are the preferential sites for reaction with an evaporated metallic atom, and if possible, to gain an idea of the relative reactivities of different chemical functions, such a progressive and systematic study was undertaken, and is still being continued in our laboratory.

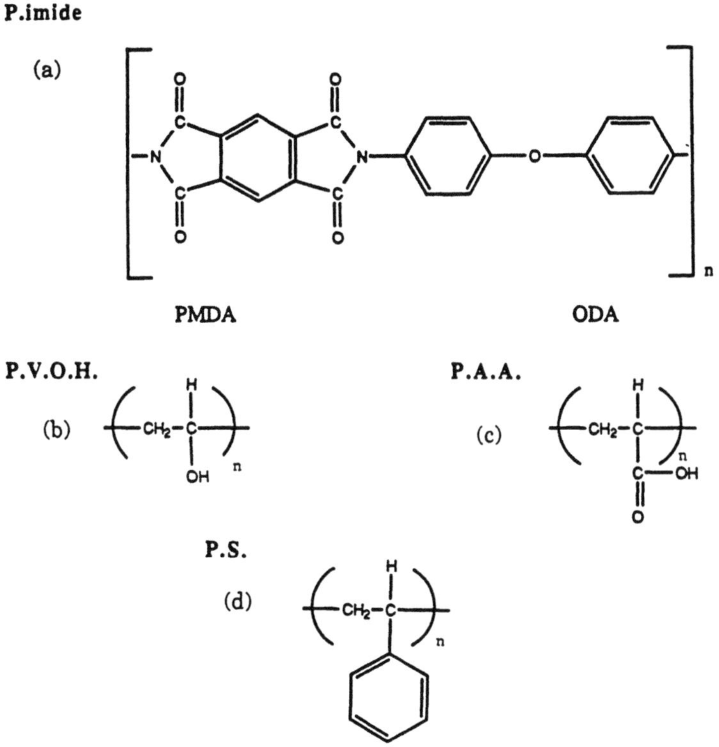

P.imide

(a)

PMDA   ODA

P.V.O.H.   **P.A.A.**

(b)   (c)

P.S.

(d)

Fig. 1: Chemical composition of the monomeric unit of the polymers selected for this study : (a) polyimide ; (b) polyvinylalcohol ; (c) polyacrylic acid ; (d) polystyrene

In order to evaluate which metal would be the most reactive when deposited on the polymer surfaces, - and thus would present the most significant HREELS fingerprints through the polymer vibrational bands - we used a thermodynamical approach, initially developed to predict the affinity of a metallic atom towards a polyimide surface, at the carbonyl site [12]. We extended these calculations, performed for silver, copper, nickel, and chromium, to other metals - which could be of technological interest - like palladium, indium, titanium and aluminum. Although this model strictly predicts only the formation of a metal oxide, and although the thermodynamical data refer to single oxides (of the type M-O) and not to natural ones (of the type M_xO_y), we present in Fig. 2 the calculated free energy of interfacial oxidation, which suggests that Ti, Cr, and Al should be the most reactive [13].

When bonding chemically onto the polymer surface, a metallic atom transfers some charge to the substrate. This ability to transfer charge is reflected in the well-known reduction standard potential scale that predicts, for reactions in solution, which chemical reactions might spontaneously take place. The electron donor or acceptor character of some metals have been compiled from the literature [14] and are

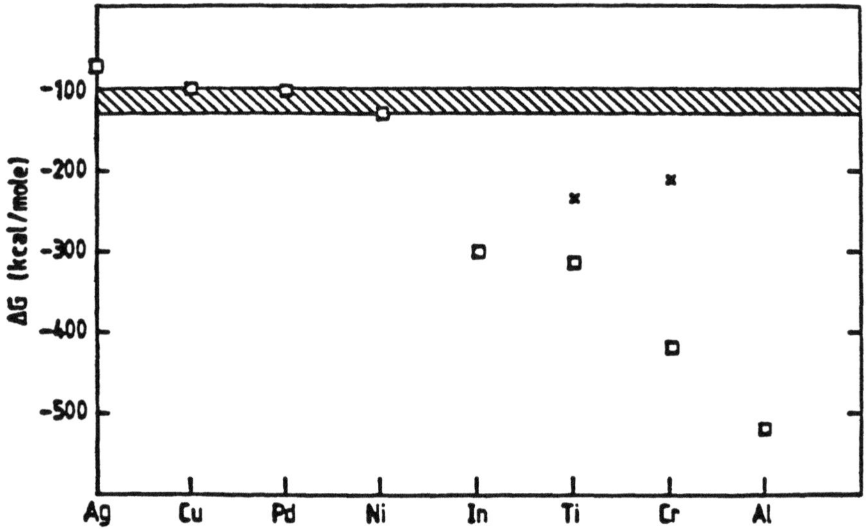

Fig.2 : Calculated free energy of interfacial oxidation of selected metals on the polyimide surface. Adapted from ref. 12. (□) M_x-O_y type ,oxide; (x) M-O type oxide.

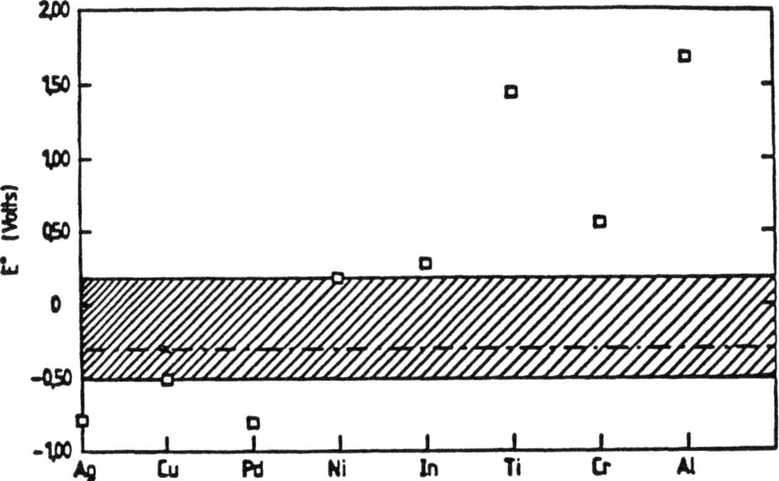

Fig. 3 : Electron donor or acceptor character of some metals, as reflected by the standard reduction potential scale (ref. 13)

presented in Fig. 3. Again, without discussing the limits of the model, these results predict a high reactivity for Ti, Cr, and Al [13].

The two thermodynamical and reduction potential models - which in fact rely on similar assumptions - agree in their conclusions. And as it is definitely easier to control the evaporation of aluminum than that of the chromium or titanium; aluminum was selected for the experiment. A UHV Knudsen cell was used to evaporate aluminum (Johnson Matthey) in situ in the submonolayer regime. A very low evaporation rate ($\simeq 1$ Å/min) could be maintained, at a base pressure of $\simeq 1 \times 10^{-10}$ Torr. Hereafter the coverages are expressed as the number of Al atoms per sample cm^2, i.e. after correction for geometrical factors taking into account the sample - quartz microbalance distance. It is useful to remember that one atomic layer of metallic aluminum contains about 1.2×10^{15} atoms/cm^2, or that a "1 Å - thick" layer on a flat surface should contain roughly 0.6×10^{15} atoms/cm^2.

4. The Aluminum-Polyimide Interface [5]

Polyimide (200 Å thick PMDA-ODA) films, spin coated onto gold decorated silicon wafers, were cured at T ~ 580K for 2h. Inside the HREEL spectrometer, they were heated again (T° $\approx$ 500 K, 1 hour) to remove surface contaminants [15]. Aluminum was evaporated stepwise from the Knudsen cell, onto the polymer at room temperature. A constant dose unit of $\approx 1.5 \cdot 10^{14}$ atoms/cm^2, i.e. about 1/10 Al monolayer, was used and cumulated on the polyimide film : selected HREEL spectra are presented in Fig. 4.

The ν(C-H), ν(C=O), ν(C-N), ν(C-O-C), and ν (OC)$_2$NC vibrational bands are among the ones identified in the electron loss spectra, and are labelled in Fig. 4a for the clean polyimide. They allow one to note that neither the (C-H) groups - from the aromatic rings -, nor the C-O-C ether linkages are affected at all by the deposition of the first Al doses (Figs. 4b and c). Conversely, the carbonyl vibrational bands are strongly reduced in intensity, but are still observable, whereas the ν(C-N) band almost disappears. New vibrational losses appear at larger Al doses (Fig. 4c), an aliphatic ν(C-H) one below 3000 cm^{-1}, a ν(OH) band at ~ 3730cm^{-1} ; finally, one broad loss feature dominates the EEL spectrum around 800 cm^{-1} : it is attributed to the formation of Al-O "oxide-like" bonds.

The partial disappearance of the C=O stretch and the parallel intensity reduction of the symmetric ν(O=C)$_2$ band localizes the interaction of the evaporated Al atoms close to this -C=O site, which indeed contains the largest dipolar moment in the PMDA-ODA monomeric unit. A rough intensity analysis of the HREELS spectra suggests that the metal atoms react with only half of the C=O sites. The simultaneous intensity decrease of the ν(C-N) band is consistent with an influence of the deposited Al atom on the neighboring C-N bond ; this confirms the idea of the existence of a large electronic delocalization in the imide entity.

For low Al coverages (Fig. 4b), no aluminum oxide species is detected in the 600-800 cm^{-1} region [16] : therefore there is no C=O bond breaking; the Al atoms approaching the C=O sites weaken but do not break the CO bonds. As a result, a red-shifted ν(C=O) band, and an blue-shifted ν(C-N) one are seen to broaden the vibrational structures

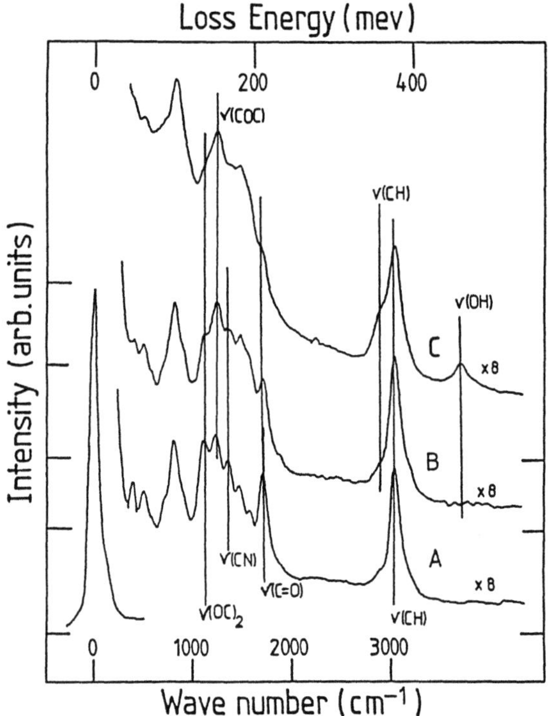

Fig. 4 : HREELS - vibrational spectra of the interface formation between a polyimide film and evaporated aluminum : (a) clean polyimide surface ; (b) with 1/10 layer of Al ; (c) with half a layer of Al

between 1450 and 1650 cm^{-1} : a -C-O-Al-like complex is formed, influencing the electron distribution within the whole imide moiety.

At larger metal coverages (Fig. 4c), new peaks at 2920 and 3730 cm^{-1} appear first, then another broad one at ~ 820 cm^{-1}. These aliphatic -CH$_n$ and OH groups on the polymer surface are probably fingerprinting residues on the surface, captured from the residual gas, or more probably released after bond scissions of the polymer skeleton. Indeed, many aluminum atoms do not react at the C=O sites - as the ν(C=O) band is still unperturbed -, but probably condense on the polymer surface in small islands, and diffuse into the polymer (as no overall HREELS intensity decrease is noticed) : this polymer degradation might thus be caused by the energy released during the Al condensation. After the equivalent of 1 to 2 layers of Al atoms on the polyimide surface, the broad and intense band at 800 cm^{-1} testifies to the existence of Al-O bonds, as in aluminum oxide.

<u>5. The Aluminum-Polyvinylalcohol Interface</u> [6]

Polyvinylalcohol (Scientific Polymer Products, Inc.) or PVOH was deposited from a water solution by spin coating on gold-covered conducting glass. No further treatment was applied to these deposited thin polymer films, which were readily transferred to the UHV chamber for the HREELS measurements.

PVOH contains a large dipolar moment on the hydroxyl group, which is readily observed on the clean polymer spectrum (Fig. 5a) at $\approx$ 3400 cm^{-1}. If one concentrates on this -OH reactive group, one should also recall the δ(C-O) deformation mode at 480 cm^{-1} as a fingerprint of this type of polymer. A more detailed analysis of the complete vibrational spectrum is presented elsewhere [6].

When a fractional Al layer is deposited stepwise from the Knudsen cell on this PVOH substrate, the induced chemical reactions at the interface are initially (for about 1/10 metal atomic layer - Fig. 5b) noticeable through the vibrational bands involving the oxygen : the ν(OH) at 3400 cm^{-1} and δ(C-O) at 480 cm^{-1} are strongly depressed, with a decrease (or blue-shift) of the ν(C-C) band at 875 cm^{-1}. After other cumulated doses of aluminum (Fig. 5c), the -OH signal is no longer discernable, while

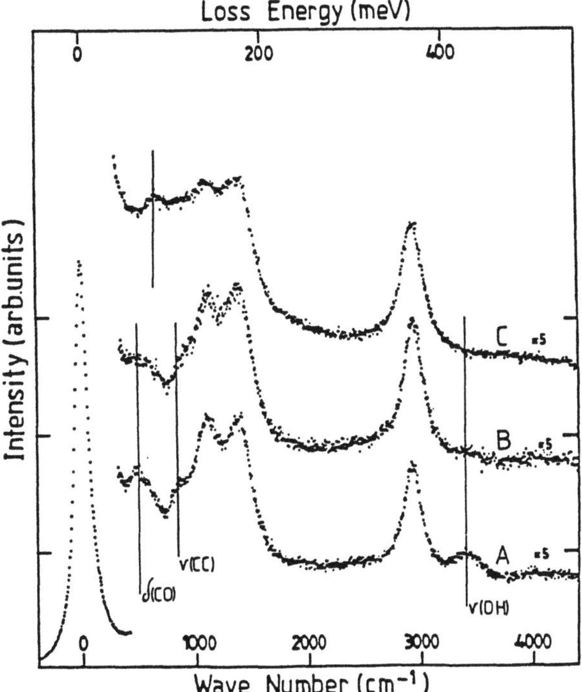

Fig. 5 : HREELS - vibrational spectra of the interface formation between a polyvinyl alcohol film and evaporated aluminum : (a) clean polyvinylalcohol surface ; (b) with1/10 layer of Al ; (c) with 1/3 layer of Al

a new vibrational band is progressively growing in intensity between 500 and 800 cm^{-1}; all the other loss bands decrease together in intensity with subsequent aluminum coverages.

The chemistry of the PVOH monomeric unit (CH$_2$-CHOH) is simpler than the one from polyimide (PMDA-ODA) : as expected, the chemical reactions during the Al deposition are more straightforward. Indeed, the Al atoms readily attack the hydroxyl site, destroying the O-H bond - as the ν(OH) band disappears - and link to the oxygen atoms. At the beginning, an Al-O-C-C- hybrid complex is formed, as the δ(O-C) as well as the ν(C-C) bands are affected. As soon as all the oxygen sites are occupied by the Al atoms, there is evidence of destruction of the polymer : the vibrational band growing below 700 cm^{-1} is characteristic of aluminum oxide (or carbide) entities, establishing C-O bond cleavages.

Finally, we would like to note that all the -OH sites are consumed during the Al deposition, even for a coverage of only 3 to 4 x 10^{14} atoms/cm^2, i.e about 1/4 or 1/3 of equivalent Al monolayer : it seems that all the Al atoms react at the hydroxyl site first, in about a 1 to 1 ratio. Further Al deposition probably condenses small clusters at these Al-O-C sites, which finally break the O-C bonds, to let Al-O type species appear.

6. The Aluminum-Polyacrylic Acid Interface [6]

Polyacrylic acid (PAA, in aqueous solution from Aldrich Chemie) was prepared as a thin film by spin coating onto gold-decorated conducting glass. After careful drying in air, the sample was introduced into the UHV chamber for the HREELS measurements. It is worthwhile mentioning that the HREELS spectra from samples gently heated in vacuum (<100°C) showed a depletion of COOH groups on the surface. Results presented here are thus from non-heated samples.

The carboxylic function (-COOH) is expected to be the most reactive entity in the PAA monomeric unit. On a clean PAA surface, this molecular group should be easily fingerprinted through its vibrational spectrum by the carbonyl stretch ν(C=O), and the hydroxyl elongation mode ν(O-H). Indeed in the HREELS data (Fig. 6a), the ν(C=O) vibration is very well resolved at 1720 cm^{-1} ; however, no peak is detected in the hydroxyl range (3000 - 3500 cm^{-1}). Instead, a very broad band (almost 1000 cm^{-1}) is observed at 2940 cm^{-1} under the ν(C-H) : this stuctureless bump is characteristic of hydrogen bonding, via the hydroxyl species ; thus, in the solid phase, PAA is seen to develop inter- and intra-chain weak hydrogen bonding [17].

Even the smallest amount of Al atoms (about 1.0 x 10^{14} atoms/cm^2, i.e. below 1/10 equivalent monolayer), induces a drastic decrease of the ν(C=O) band intensity, and a modification of the ν(O-H) bump as a structured band appears now at about 2450cm^{-1} (this is not shown in the figure). And already at 1/10 Al coverage, the ν(C=O) band is no longer distinguishable (fig. 6b) : instead some loss intensity has been added at ~

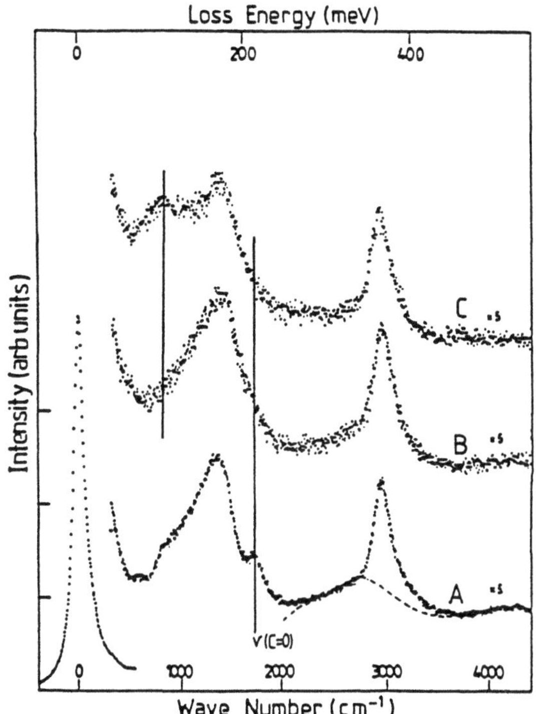

Fig. 6 : HREELS - vibrational spectra of the interface formation between a polyacrylic acid film and evaporated aluminum :(a) clean polyacrylic surface ; (b) with 1/10 layer of Al ; (c) with 1/2 layer of Al

1500 cm^{-1}, and around 700 cm^{-1}. With further aluminum deposition, a well-resolved band grows at 800 cm^{-1}, whereas more loss intensity is added between 2000 and 2400 cm^{-1} (Fig. 6c : 1/2 Al monolayer).

This study of the incipient interface formation between evaporated aluminum atoms and a polyacrylic acid film confirms that the metal is fixed close to the very reactive carboxylate function : the disappearance of the C=O stretch band and the increase of loss intensity between 1450 and 1700 cm^{-1} is compatible with the formation of a carboxylate ion -$CO_2^-Al^+$ whose vibrational fingerprint is expected at about 1600 cm^{-1} [18]. Simultaneously, hydrogen has to be deplaced ; therefore the shape of the ν(OH) bump is modified. The breaking of the hydrogen bonding within the polymer film opens therefore a new reaction scheme, where the aluminum atoms bond to the oxygen ones and form Al-O species : this is testified by the intensity increase between 600 and 800 cm^{-1} [16]. An alternate explanation to the band appearing at 600 cm^{-1} (Fig. 6b) is the possible excitation of a rocking (CO_2^-) mode, as measured at 565 cm^{-1} by IETS for PAA adsorbed on aluminum oxide [19] ; this vibrational band should thus be the natural companion to the ν(CO_2^-) carboxylate one at about 1600 cm^{-1}. At higher Al coverages, the band growing at $820 cm^{-1}$ is to be associated with Al-O or O-Al-O oxide-like species.

The reaction of Al atoms on the carboxylic sites is immediate and total, already for a coverage of 1.5×10^{14} atoms/cm^2 (Fig. 6b), i.e. an equivalent of about 1/10 of layer of Al atoms. This suggests a 1 to 1 reaction of the metal atom versus the polymer surface monomeric unit.

7. The Aluminum-Polystyrene Interface

Polystyrene thin films were prepared by dipping a gold-coated glass slide into a solution with carbon tetrachloride. They were analyzed without any further treatment. Aluminum was dosed with the same Knudsen cell as for the previous experiments. But the first evaporations did not induce any modification at all of the HREEL spectra. Up to a coverage of 1×10^{15} atoms/cm^2, i.e. about one equivalent aluminum layer, no attenuation of any band was recorded, but the elastic peak was slightly broadened. This effect is progressively enhanced up to a coverage of about 10 layers of aluminum, when suddenly drastic modifications of the vibrational bands are evident : the intensity of the ν(C-H) aromatic stretch is then about half attenuated, and a new band develops at ~ 820 cm^{-1}. Another large Al dose (6×10^{15} atoms/cm^2) strongly decreases the intensity of the ν(C-H) and δ(CH) bands associated with the unsaturated carbon-hydrogen bonds. Subsequent evaporations attenuated all the bands, in favor of the 820 cm^{-1} one. It is to be noticed, however, that a signal from the C-H aliphatic and aromatic stretches was still detected for Al coverages larger than 20 equivalent layers.

Polystyrene does not offer any particular site for the Al atom condensation : the pendant-phenyl rings do not seem very attractive as many layers of aluminum are deposited without any visible interaction. As the HREELS spectra are not attenuated by these evaporations, it is thought that the Al atoms diffuse into the polymer film - even at room temperature. After a coverage of about 10 Al layers, about half of the aromatic C-H stretch vibrations are damped : the Al atoms are finally interacting with these delocalized electronic systems, and the band growing at 820 cm^{-1} is to be attributed to an Al-C stretch, as in organo-aluminum compounds [20]. The subsequent evaporations continue to cover the polymer surface which is still visible : Al islands, which nucleated around some aromatic rings, continue to develop extremely slowly.

The affinity of Al atoms for the polystyrene surface is - according to these HREELS experiments - very poor : Al atoms condense and diffuse on and through the polymer surface before finally localizing on some of the benzene rings. Aluminum-carbon bonds are then formed. Metal clusters grow progressively on the polymer surface.

<u>8. Comparison of the XPS - HREELS Experiments</u>

The set of HREELS data presented here, and the XPS results briefly summarized at the beginning of this presentation, prove that both types of experiments are really complementary. For the same phenomena - the chemical bonding of a metal atom to a polymer chain - the two spectroscopies use different analyses, a core level chemical shift description of the electronic structure (XPS), and a charge transfer analysis of the vibrational band intensities and shifts, respectively.

As in XPS [8,9], a significant reactivity difference is noted between polymers containing no oxygen, single bonded (OH) and double bonded oxygen atoms (C=O) ; the case of polyacrylic acid appears particularly complex [6]. The ability of HREELS to detect hydrogen (via e.g. the v(C-H) and v(O-H) vibrational bands), is a significant advantage in detecting hydrogen bonding effects.

As long as there is no theoretical background to support a quantitative analysis of the relative intensities of different vibrational bands, HREELS cannot be used as ESCA to quantify the relative number of various chemical groups on the polymer surface [3]. However, HREEL data can be used in dynamical experiments - to evidence redistributions of molecular groups on the polymer surface and show, e.g. that heated PAA contains fewer carboxylic groups than expected on its surface.

The present HREELS data could not positively identify or refute the formation of metal carbide species on the metallized polymers : the experimental resolution is presently not good enough to separate Al-C and Al-O stretch vibrations.

It is expected that high resolution electron energy loss spectroscopy will yield information on the molecular configuration of polymer-metal complexes. Such a type of study, based on angular resolved experiments, has been extremely successful in elucidating part of the geometrical configuration of the clean polyimide surface [5]. For the metallized polymers however, all the data recorded up to now are just consistent with a statistically averaged orientation of the different molecular groups. But this result might just be an artefact, caused by the roughening of the polymer surface during the metallization : further work on this subject is in progress.

<u>9. Acknowledgements</u>

This work has been financially supported by the Belgian National Fund for Scientific Research, the IRSIA, the Belgian program on Interuniversity Attraction Poles (Prime Minister's Office - Science Policy Programming), a Du Pont de Nemours donation, and a CEE-BRITE contract # RI-1B-0178.

10. REFERENCES

1. H. Ibach and D.L. Mills, <u>Electron Energy Loss Spectroscopy and Surface Vibrations.</u> (Academic, New York 1982).

2. P.A.Thiry, M. Liehr, J.J. Pireaux, and R. Caudano, Phys. Scr. 35, 368 (1987).

3. J.J. Pireaux, C. Grégoire, M. Vermeersch, P.A. Thiry, and R. Caudano, Surf. Sci. 189/190, 903 (1987).

4. N.J. Dinardo, J.E. Demuth, and T.C. Clarke, Chem. Phys. Lett. 121, 239 (1985) ; J. Vac. Sci. Technol. A4, 1050 (1986).

5. J.J. Pireaux, M. Vermeersch, C. Grégoire, P.A. Thiry, R.Caudano, and T.C. Clarke, J. Chem. Phys. 88, 3353 (1988)

6. Y. Novis, et al., submitted for publication ; N. Degosserie, Facultés Universitaires de Namur, Mémoire de Licence (1988), unpublished.

7. J.J. Pireaux et al., in preparation.

8. J.M. Burkstrand, J. Appl. Phys. 52, 4795 (1981) ; and references therein.

9. B.M. Dekoven, and P.L. Hagans, Appl. Surf. Sci. 27, 199 (1986)

10. L.J. Atanasoska, S.G. Anderson, H.M. Meyer, Z. Lin, and J.H. Weaver, J. Vac. Sci Technol. A5, 3325 (1987)

11. M. Liehr, P.A. Thiry, J.J. Pireaux, and R. Caudano, Phys. Rev. B 33, 5682 (1986).

12. N.J. Chou and C.H. Tang, J. Vac. Sci. Technol. A2,751 (1984).

13. M. Vermeersch. Facultés Universitaires de Namur, Mémoire de Licence (1987), unpublished.

14. CRC Handbook of Chemistry and Physics.

15. J.J. Pireaux, C. Grégoire, P.A Thiry, R. Caudano, and T.C. Clarke J. Vac. S Technol. A5, 598 (1987).

16. M. Liehr, P.A. Thiry, J.J. Pireaux, and R. Caudano . J. Vac. Sci. Technol. A2, 1079 (1984).

17. F. Laborie, J. Polym. Sci. 15, 1255 (1977).

18. For example : A. Ruaudel-Teixier and A. Barraud, J. Chimie Physique 84, 469 (1987).

19. R.F. Colletti, H.S. Gold, and C. Dybowski. Appl. Spectrosc. 41, 1185 (1987).

20. T. Moole and E.R. Jeffery, <u>Organoaluminum Components</u> (Elsevier, Amsterdam 1972).

Adsorption Structure of Organic Molecules on Mo(110)

C.M. Friend

Department of Chemistry, Harvard University,
12 Oxford St., Cambridge, MA 02138, USA

1. Introduction

The stability of a metal-polymer interface is dependent upon the strength of interaction between functional groups in the polymer and the stability of the interface with respect to degrading reactions. A knowledge of the adsorption structure of surface species that serve as models for polymer subunits or monomers is central to understanding their stability and bonding to the surface. The interplay between structure and intermolecular interactions is of particular importance and only recently have tools become available to probe surface structure. Several methods have been developed to probe surface structure, and in particular, molecular orientation. The most widely used of these methods are low energy electron diffraction (LEED) [1], angle resolved photoelectron spectroscopy[2], and more recently, near edge X-ray absorption fine structure (NEXAFS) measurements[3].

In order to determine the structure of an adsorbed species, it is necessary to measure reactivity and to use other spectroscopic methods to identify surface intermediates. To then relate the structure to adhesion properties from a microscopic perspective, the reactivity of surface intermediates must be measured. The focus of this article is the application of the NEXAFS method to determine adsorption structure in conjunction with X-ray photoelectron spectroscopy, high resolution electron energy loss spectroscopy and temperature programmed reaction spectroscopy which are used to identify intermediates and measure reactivity.

2. The NEXAFS Method

The near edge X-ray absorption fine structure (NEXAFS) technique has been recently developed as a probe of orientation of adsorbed species[4]. The strength of the method is that structural information can be obtained for relatively complex molecules. Long range translational order is not required for determination of molecular orientation using the NEXAFS method, a significant advantage over LEED. In order to probe the molecular geometry of a surface intermediate by any method, a single molecular intermediate must be isolable on the surface. If more than one molecular intermediate is present on the surface, an orientation will be derived, and in general, the contributions from the different intermediates cannot be deconvoluted when a mixture of surface species is present. In the case of NEXAFS, there are usually not significant chemical shifts associated with specific intermediates making spectral deconvolution virtually impossible. In NEXAFS, the contribution of atomic adsorbates can sometimes be subtracted as for benzyne formed from benzene on Mo(110). The requirement for isolation of a single molecular intermediate on the surface in order to determine structure mandates the use of complementary spectroscopic probes. In our work, X-ray photoelectron and high resolution electron energy loss (vibrational) spectroscopies have been chosen to identify the coverage and temperature necessary for isolation of intermediates.

Springer Series in Surface Sciences, Vol. 17
Adhesion and Friction   Editors: M. Grunze and H.J. Kreuzer
© Springer-Verlag Berlin, Heidelberg 1989

In the NEXAFS experiment, tunable, linearly polarized light ($\sim$85 % polarization) induces transitions from core levels to unfilled bound and continuum states. Thus, synchrotron radiation is required for the performance of NEXAFS experiments. The data shown here were obtained at the Stanford Synchrotron Radiation Laboratory. The electronic transitions are induced as the photon energy is swept through the X-ray absorption threshold for the core level of interest. In the so-called near edge region, corresponding to 0-50 [eV] above the absorption threshold, multiple scattering events predominate. For excitations of low Z atoms in the adsorbed intermediate, the transitions can be formally thought of as being transitions to molecular states that are antibonding and unfilled in the ground state. The dipole selection rule can be applied, and the molecular orientation of an adsorbed intermediate determined from the polarization dependence of the transitions of different symmetry.

The utility of NEXAFS for structural studies of both simple and complex molecules on surfaces is well established. Stohr and Jaeger first utilized the method to determine the orientation and bond length of CO adsorbed on Ni(100)[4]. Since then the orientations of a variety of more complex organic intermediates, both with and without C-C bond unsaturation, have been studied on several transition metal surfaces, including Mo, Pt, Ag, W and Ni. For example, the ring orientation of the aromatic hydrocarbons, thiophene[5], pyridine[6] and benzene[7,8] have been derived from NEXAFS measurements. The orientation of surface species lacking π-bonds has also been extracted from NEXAFS experiments, including methyl thiolate adsorbed on Pt(111)[9] and methoxy groups adsorbed on Cu(100)[10].

The mathematical details of the method are described in detail elsewhere[3]. We will qualitatively discuss the example of benzene coordinated to a metal surface for illustrative purposes here. In gas phase and condensed benzene, both π^* and σ^* resonances are observed and have been assigned with the aid of calculations. The initial state is the totally symmetric C(1s) state and, therefore, the resonances of π and σ symmetry will have opposite polarization dependences. The chemisorbed benzene molecule will generally orient itself with respect to the surface and, therefore, the polarization dependence of the resonances can be determined by measuring their intensity variations as a function of the angle of photon incidence. Since synchrotron radiation is linearly polarized, the electric field vector is perpendicular to the direction of light propagation. The π^* resonances will have maximum intensity when the electric field vector is oriented perpendicular to the plane of the benzene ring while the σ^* resonances will have maximum intensity when $\vec{E}$ is oriented in the plane of the ring. Therefore, for benzene bound parallel to the surface plane, the π^* resonances have maximum intensity for glancing photon incidence while the σ^* resonances have maximum intensity at normal photon incidence. The geometry of the NEXAFS experiment is shown in Fig. 1.

There are several complications to the interpretation of NEXAFS data that are not discussed in detail here, but should be mentioned. Upon interaction with the surface, the σ^* resonances of molecular benzene may be shifted somewhat in energy and the π^* states may be partially occupied due to donation of electron density from the d-band. For the case of strong bonding with the metal, additional resonances due to the formation of new orbitals between the metal and carbon framework may appear[11]. This complicates the intensity analysis because the new resonances due to strong C-metal bonding may be in the same energy range as the π^* transitions. Furthermore, symmetry breaking due to interaction with the metal surface may render analysis of the spectra difficult. These details of spectral analysis are the subject of current investigation.

An important feature of NEXAFS spectra is that the spectral features of specific functional groups remain essentially unchanged in different molecules[12]. Thus, the spectra for larger molecules can be analyzed in terms of the functional groups present. For example, in this work, the structure of the phenyl thiolate ($-S-(C_6H_5)$) intermediate on Mo(110) can be

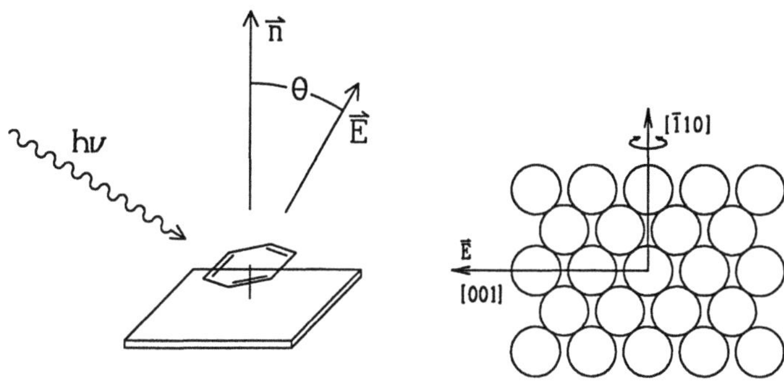

Figure 1. The geometry of the NEXAFS experiments with respect to the Mo(110) surface.

analyzed readily since the phenyl group will produce resonances very similar to benzene. This is a simplifying feature in the analysis of complex adsorbed intermediates that may not have gas phase analogues or reference spectra.

3. Other Spectroscopic Probes

Complementary spectroscopic probes are necessary for identification and isolation of molecular intermediates on the surface. As emphasized above, it is necessary to know the identity of surface intermediates and to isolate them for a structural determination with any currently available structural probe. Identification of the intermediates on the surface is taken for granted, but is sometimes difficult since the possibility of bond breaking and formation must be taken into account even at moderate temperatures. An early example of a case where such reactive events were not considered in LEED analysis is that of ethylene and acetylene on Pt(111). Originally, structures were proposed for the intact molecules but the intermediate present is actually ethylidyne, $CC(H_3)$, as was ultimately born out in more refined LEED calculations[13].

In this work, we utilize three methods to identify surface intermediates: temperature programmed reaction, X-ray photoelectron and high resolution electron energy loss spectroscopies. This combination is by no means unique, but is a good complement for determining the identity of surface intermediates. Importantly, none of these methods is capable of identifying surface intermediates as "stand alone" techniques. The full complement of methods is necessary for molecules of the level of complexity described here. All of these methods have been used extensively in surface analysis, so that a detailed discussion of each technique is unwarranted. Rather, a brief description of each as applied to this work is given below.

Temperature programmed reaction spectroscopy is a method for probing reaction and desorption kinetics of adsorbed species[14]. When used in conjunction with selective isotopic labelling or isotopic exchange, the composition of surface intermediates that give rise to volatile products can be determined. This method is capable of monitoring subtle chemical phenomena which may not be observable spectroscopically. For example, rapid reversible C-H bond breaking may occur for adsorbed species. In this work, temperature programmed reaction was necessary for identification of surface benzyne, an intermediate formed in the reaction of benzenethiol on Mo(110). It was also used to probe the relative kinetics for O-H and C-H bond scission in phenol on Mo(110). For complex molecules that undergo surface reaction, temperature programmed reaction spectroscopy is an essential probe.

X-ray photoelectron spectroscopy is a means of empirically identifying different chemical species on the surface based on chemical shifts in core level binding energies[15]. X-ray photoelectron spectroscopy only probes intermediates that remain on the surface and is sensitive to irreversible transformations. In our work, X-ray photoelectron spectroscopy is used to determine the conditions of coverage and temperature where a single molecular intermediate is present on the surface, to determine the temperature where bond breaking occurs as an approximate measure of kinetic stability, and to aid in the identification of surface intermediates. X-ray photoelectron spectroscopy is sensitive to irreversible bond breaking and formation process, in particular where a heteroatom such as oxygen is present, since the chemical shifts in the core level binding energies of electronegative atoms are >0.5 [eV]. The C(1s) binding energies of carbon centers bound directly to a heteroatom are also shifted by as much as 1.9 [eV]. Chemical shifts in C(1s) binding energies of adsorbed hydrocarbons not containing heteroatoms are considerably smaller, but atomic carbon can readily be distinguished from species containing intact C-H and C-C bonds.

Vibrational spectroscopy is necessary to unequivocally identify surface intermediates. The chemical shift of a core level binding energy in X-ray photoelectron spectroscopy cannot be uniquely assigned to a specific intermediate, but is assigned by comparison to the binding energies of known compounds. Vibrational spectroscopy is sensitive to the specific molecular arrangement in the adsorbed species and is, therefore, complementary to both X-ray photoelectron and temperature programmed reaction spectroscopies. In this work, high resolution electron energy loss spectroscopy has been used as a vibrational probe[16]. High resolution electron energy loss spectroscopy has a high sensitivity and is not governed by rigid selection rules. While the latter renders determination of molecular orientation using high resolution electron energy loss spectroscopy exceedingly difficult, it has the advantage that the absence of specific modes in the vibrational spectrum can be indicative of a specific type of bond cleavage. In particular, modes involving motion of hydrogen atoms have a large component of "impact scattering" so that X-H (X=C,S,O) stretch modes are observable even if oriented parallel to the surface plane[16]. This is important in the identification of the phenyl thiolate intermediate formed from S-H bond dissociation in benzenethiol upon adsorption on Mo(110).

4. The Reactivity and Structure of Benzene on Mo(110)

Benzene was investigated due to its importance in heterogeneous catalysis and because it is important in assessing the relative contributions of the π system of the aromatic ring and σ-donation of heteroatoms in molecules, such as phenol, aniline or benzenethiol, that serve as good models for polymer precursors or lubricants. Benzene, containing no heteroatom, is a measure of the effect of the aromatic ring upon adsorption and reactivity.

Benzene has been widely studied on transition metal surfaces, including Ni[17], Pd[18,19], Pt[20], Cu[21], Ag[22], Rh[23] and Ru[24]. Benzene is generally found to adsorb molecularly at temperatures below 300[K] with the ring oriented parallel to the surface. The possibility of dissociative chemisorption at low temperature has been suggested on Ru(0001)[24] and recently, a nonparallel geometry has been suggested for benzene bound to the highly corrugated Pd(110) surface.[19] The structural determinations on these transition metal surfaces was made using a variety of methods including high resolution electron energy loss spectroscopy, angle resolved ultraviolet photoemission spectroscopy, LEED and NEXAFS.

The observed predominance of parallel bonding to the transition metal surface is consistent with bonding of benzenes to transition metal complexes where d$\rightarrow\pi^*$ and $\pi\rightarrow$d bonding can occur[25]. Parallel geometry is expected to maximize overlap between the d orbitals of the metal and the π orbitals of the benzene on a transition metal surface. Indeed, cluster calculations modeling benzene bound to Ni(111) support this picture[26].

As on other surfaces, benzene is found to bind parallel to the Mo(110) plane when molecularly bound. The experimental details of this work are described elsewhere[8]. The first task in understanding the bonding and structure of benzene on Mo(110) is to determine the conditions where molecular adsorption predominates. Temperature programmed reaction and X-ray photoelectron spectroscopies were used to probe the stability of molecular benzene on Mo(110).

Benzene undergoes competing molecular desorption and irreversible decomposition near 350 [K] on Mo(110). Temperature programmed reaction data for benzene on Mo(110) after adsorption of multilayers is shown in Fig. 2. Approximately 10% of the chemisorbed benzene desorbs molecularly while the remaining decomposes via both C-H and C-C bond breaking steps in the range of 350-550 [K]. Irreversible decomposition yields surface carbon and gaseous dihydrogen. No other gas phase products are observed during temperature programmed reaction. The three peaks labelled a_1, a_2, and a_3 below 200 [K] are characteristic of benzene weakly interacting with the chemisorbed layer and are observed on several other surfaces, including Ni(100)[17], Ru(0001)[24] and Rh(111)[27]. These physisorbed states will not be discussed further.

X-ray photoelectron data for benzene chemisorbed on Mo(110) below 300 [K] are consistent with the presence of intact molecular benzene. C(1s) X-ray photoelectron data for benzene monolayers are shown in Fig. 3. A single resolvable feature with a binding energy of 284.0 [eV] is observed. The width of the C(1s) peak for the molecular species is 1.2 [eV], consistent with the presence of a single intermediate on the surface. The binding energy of 284 [eV] is typical of hydrocarbons adsorbed on transition metal surfaces and is consistent with, but not unambiguous evidence for, the presence of molecular benzene on the surface. Therefore, additional evidence for intact benzene must be obtained other than X-ray photoelectron spectroscopy.

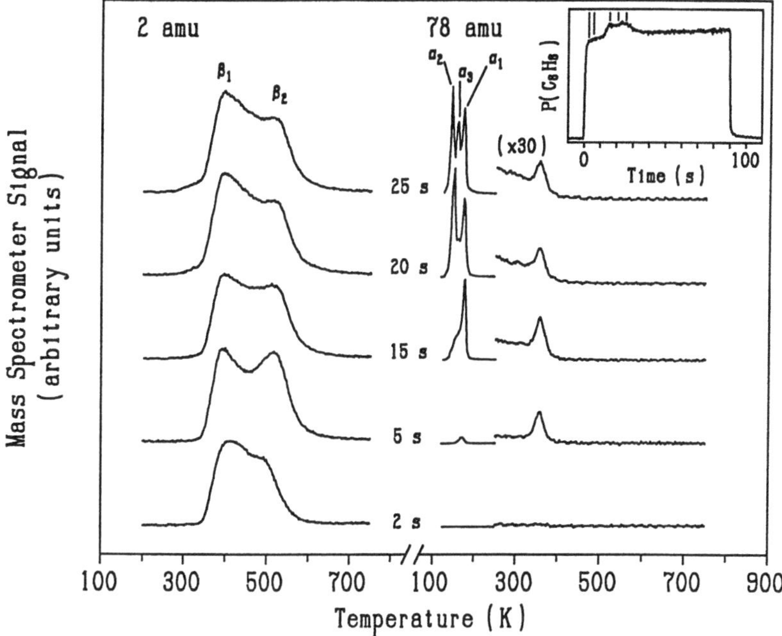

Figure 2. Temperature programmed reaction spectra of benzene on Mo(110) taken from Reference 8.

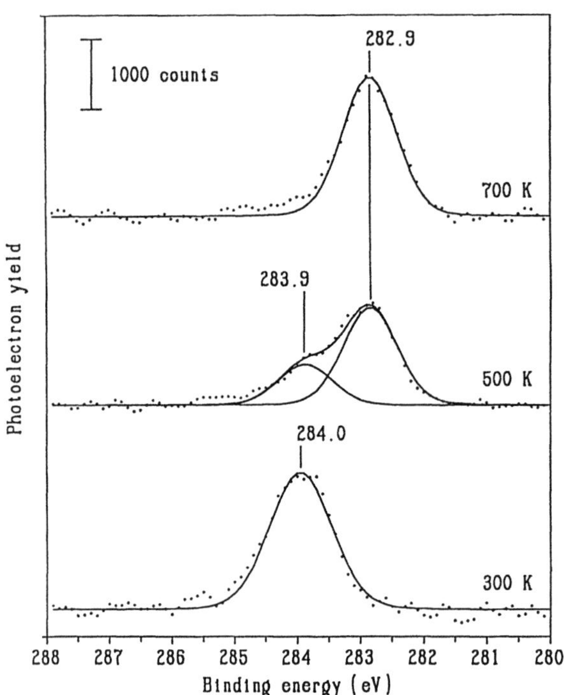

Figure 3. C(1s) X-ray photoelectron spectrum for benzene on Mo(110) taken from Reference 8. The spectra were obtained by annealing multilayers to the temperature indicated.

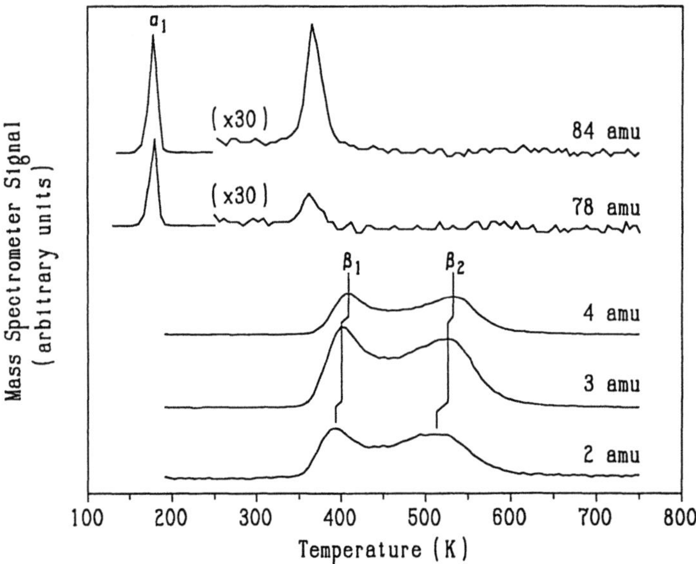

Figure 4. Temperature programmed reaction of a mixture of C_6D_6 and C_6H_6 on Mo(110) taken from Reference 8.

Isotopic exchange experiments provide confirming evidence that benzene is bound molecularly to the Mo(110) surface up to the onset of desorption at 300 [K]. Temperature programmed reaction of a ~1:1 mixture of C_6H_6 and C_6D_6 exhibit an isotope effect in the rate of benzene desorption and dihydrogen formation as shown in Fig. 4. The peak temperature and the relative amount of C_6H_6 molecular desorption are lower than those for C_6D_6. A leading edge analysis of the desorption peak yields an enthalpy of desorption of 22 [kcal/mol]. Furthermore, no isotopic exchange in the benzene desorbed is detected; only C_6H_6 and C_6D_6 are desorbed. In addition, the peak temperatures for the three dihydrogen isotopes in the lowest temperature β_1 peak are different, with the D_2 peak being 20 [K] higher than the β_1-H_2 peak. All of these observations demonstrate that benzene is molecularly bound up to the onset of desorption. The differences in the dihydrogen peak temperatures and the amount of desorption of C_6H_6 and C_6D_6 are explained by recognizing that the rate of benzene desorption ultimately drops to zero because of the competing irreversible decomposition which involves C-H(D) bond breaking. Because the rate of C-H bond cleavage is faster than C-D bond breaking, irreversible decomposition of h_6-benzene dominates at a lower temperature than for the d_6-benzene. X-ray photoelectron data obtained after heating to temperatures in the range of 350-550 [K] show that atomic carbon is present along with a molecular fragment with respective C(1s) binding energies of 282.9 and 284.1 [eV]. The molecular fragment present on the surface is identified as chemisorbed benzyne (C_6H_4) based on NEXAFS measurements, as described in detail elsewhere [8].

A parallel bonding geometry was determined for the orientation of the benzene ring in the chemisorbed state on Mo(110) using the NEXAFS method[8]. NEXAFS spectra for benzene multilayers annealed to 200[K] are shown in Fig. 5. There are strong interactions between the benzene ring and the Mo surface which leads to an apparent broadening of the

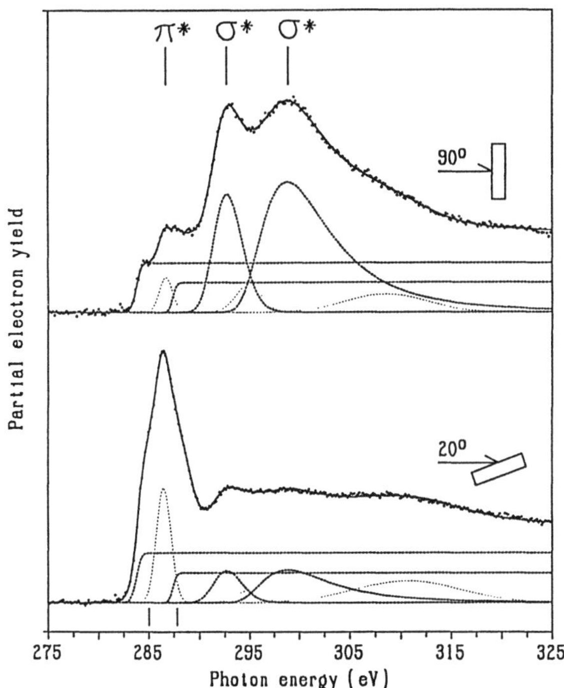

Figure 5. NEXAFS spectra for benzene chemisorbed on Mo(110). The data was obtained after heating benzene mulitlayers to 200 [K].

π^* resonance near 285 [eV]. This apparent broadening is attributed to several new resonances due to the presence of strong C-Mo bonding and render a reliable intensity analysis of the π^* resonance difficult. Therefore, the σ^* resonances at 292.7 and 298.9 [eV] are used to determine the orientation. An intensity analysis of these peaks is consistent with a nearly parallel ring orientation with a maximum possible tilt angle of 5°[8].

The complication of new resonances in the π^* region may be expected for other strongly interacting systems and is currently the subject of study in our group. Such an apparent broadening is not observed on more weakly interacting surfaces such as Cu(110) and Ag(110)[11]. Specifically, possible lowering of the adsorbate symmetry and the appearance of new resonances due to formation of M-C σ^* orbitals may give rise to complications in the interpretation of NEXAFS data. We note that these same considerations will apply in angle resolved ultraviolet photoemission spectroscopy.

5. <u>The Phenyl Thiolate Intermediate at Saturation Coverage</u>

Benzenethiol was investigated as part of a broad study of the desulfurization of organic thiols on Mo(110)[28]. This class of compounds is also of importance in lubrication since long-chain thiols are being studied for their strong binding and self-assembly on Au and other weakly interacting surfaces[29]. Due to strong coverage effects in the reactivity and bonding of the phenyl thiolate intermediate formed from benzenethiol on Mo(110), the limit of saturation coverage is discussed first.

Benzenethiol forms phenyl thiolate upon adsorption at 120 [K] on Mo(110)[28]. The reaction scheme for benzenethiol at saturation coverage on Mo(110) is represented in Fig. 6. Saturation coverage is determined to correspond to 0.12 thiolates per Mo surface atom by measuring the S(2p) integrated intensities at saturation and comparing to the corresponding intensities of the Mo(110)-p(2x2)-S surface where the S coverage is known to be 0.25.

Selective reaction of the phenyl thiolate occurs on initially clean Mo(110) at saturation coverage: the presence of surface sulfur is not required. During temperature programmed reaction, the phenyl thiolate reacts by two competitive pathways. The first pathway, involving hydrogenolysis of the C-S bond, leads to the formation and immediate evolution of C_6H_6 in the range of 250-550 [K]. Integrated X-ray photoelectron spectra show that $\approx 40\%$ of the phenyl thiolate reacts to form C_6H_6. Competing with C_6H_6 formation is the conversion of phenyl thiolate to surface hydrogen, surface sulfur, and chemisorbed benzyne (C_6H_4). Chemisorbed benzyne is exceedingly stable on Mo(110), not decomposing to β_2-H_2 until 680 [K], corresponding to a reaction of barrier 41 [kcal/mol]. Benzyne formation is estimated to occur at approximately 370 [K], based on X-ray photoelectron spectroscopy. Surface hydrogen produced during benzyne formation that does not react with phenyl thiolate undergoes atom-atom combination at 370 [K] to form β_1-H_2.

The formation of the phenyl thiolate at 120 [K] is demonstrated by both X-ray photoelectron and high resolution electron energy loss spectroscopies. The X-ray photoelectron and high resolution electron energy loss spectra are shown in Fig. 7 and 8 respectively. The S(2p) binding energies for the chemisorbed monolayer are shifted by 0.5 [eV] to lower binding energy compared to the condensed thiol, indicating a change in the environment of the sulfur upon adsorption. Such a shift could be due to S-H bond breaking or to increased core hole screening for a thiol bound directly to the metal surface. The C(1s) X-ray photoelectron data are consistent with a phenyl ring where one carbon remains bound to the sulfur center. The C(1s) peak for the chemisorbed species is similar in shape to the condensed thiol, having an asymmetry on the high binding energy side, which is optimally fit by two peaks with binding energies of 285.3 and 284.0 [eV] with relative intensities of 1 and 5, respectively.

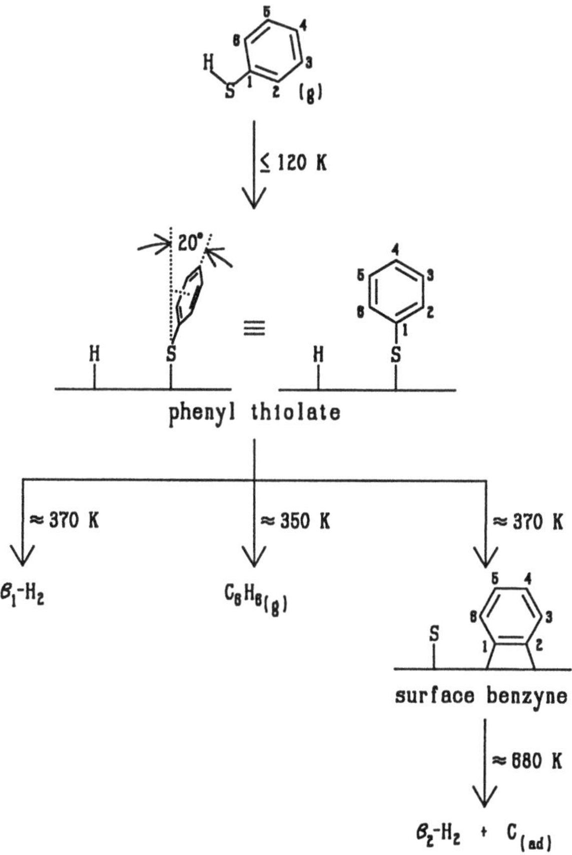

Figure 6: The reaction scheme for benzene thiol at saturation coverage taken from Reference 28.

The vibrational data demonstrates that the small S(2p) chemical shift is due to S-H bond dissociation to produce the phenyl thiolate upon adsorption. After adsorption of less than a saturation amount of benzenethiol on Mo(110) at 120 [K], the ν(S-H) mode, observed at 2590 [cm^{-1}] in condensed benzenethiol, is absent. All other modes are still present in the adsorbed species and are not dramatically shifted with respect to benzenethiol multilayers. Notably, modes associated with the phenyl ring are still present. Assignments were made in analogy to the solid state infrared spectrum and are given in Fig. 8 The absence of the ν(S-H) mode is a reliable probe of S-H bond scission since the mode should have a large component of impact scattering and should be observable even if the bond vector were oriented nearly parallel to the surface plane. All other modes are still present and the ν(C-H) mode is unshifted with respect to the benzenethiol multilayers.

Temperature programmed reaction spectroscopy lends supporting evidence to the presence of the phenyl thiolate intermediate on the surface and specifically rules out the possibility of reversible C-H bond scission upon adsorption. As shown in the scheme above, gaseous benzene is formed in the temperature programmed reaction of benzenethiol in the range of 250-550[K]. Reaction of the benzenethiol in the presence of a saturation coverage of deuterium atoms yields only d$_1$- and d$_0$-benzene, demonstrating that only one C-H bond is

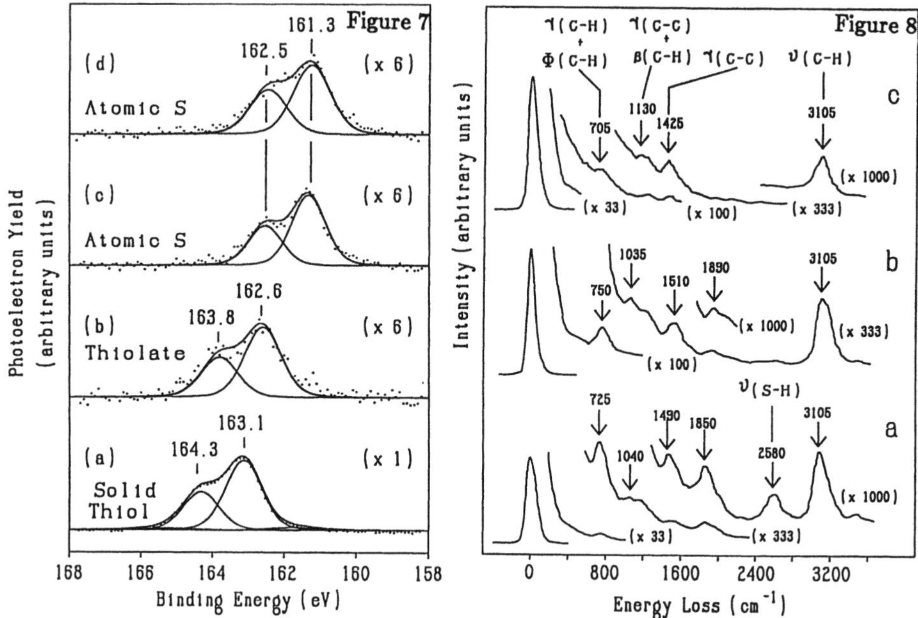

Figure 7. X-ray photoelectron spectra for intermediates formed from benzenethiol reaction on Mo(110) at saturation coverage.

Figure 8: High resolution electron energy loss spectra for intermediates formed from benzenethiol reaction on Mo(110): a.) benzenethiol multilayers at 120 [K]; b.) phenyl thiolate present after a saturation exposure of benzenethiol at 120 [K]; and, c.) benzyne formed by annealing multilayers of benzenethiol to 550 [K].

made during benzene formation. This is consistent with benzene formation proceeding through a single hydrogenolysis of the phenyl thiolate. Reversible C-H bond breaking and formation is ruled out.

The orientation of the phenyl ring in the thiolate intermediate was determined at saturation coverage using the NEXAFS method. As is evident from inspection of the intensities of the σ^* resonances shown in Fig. 9 the phenyl ring is clearly not parallel to the surface plane, but rather more nearly perpendicular. The σ^* resonances are more intense at glancing than normal photon incidence, the opposite that predicted for parallel geometry and as is observed for benzene. A curve fitting analysis renders an apparent ring tilt of 20° with respect to the surface normal with no azimuthal ordering as shown in the inset of Fig. 9[3]. The origin of the apparent ring tilt is probably due to low frequency bending vibrations of the adsorbed thiolate, since it is a non-chemical bond angle. Since the electronic transitions probed by the NEXAFS measurements are fast compared to typical vibrational lifetimes, a distribution of ring orientations would be probed if vibrational excursions are significant, resulting in the apparent tilt. An analogous effect has been proposed for formate on Ag(110) where both NEXAFS and ESDIAD measurements are consistent with low frequency vibration[30]. Comparison with organometallic analogs suggests that perpendicular bonding should be favored although a tilt of greater than 60° would be plausible if d-π bonding were important.

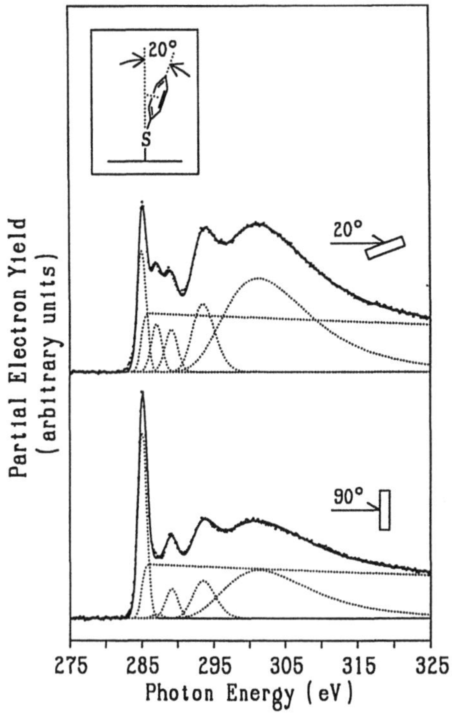

Figure 9. Near edge X-ray absorption fine structure (NEXAFS) data for phenyl thiolate on Mo(110) at saturation coverage. The data were obtained after annealing benzenethiol multilayers to 200 [K]. The inset shows the structure derived from the data.

The nearly perpendicular orientation of the phenyl ring in the thiolate intermediate is consistent with the selective formation of benzene and benzyne from its reaction at high coverage. A space-filling model of the phenyl thiolate at high coverage is shown in Fig. 10. It is clear from the model that only the ortho C-H bonds are sterically accessible to the surface. Thus, ortho-dehydrogenation should be favored which would lead to selective benzyne formation, consistent with observation.

Surface benzyne is an unprecedented intermediate, but its characterization by temperature programmed reaction, high resolution electron energy loss, X-ray photoelectron, and NEXAFS spectroscopies excludes the formation of any other intermediate. As was true of surface phenyl thiolate, temperature programmed reaction spectroscopy excludes the formation of an otherwise plausible surface species, a chemisorbed phenyl group (C_6H_5). The relative intensities of the β_1 and β_2 peaks were measured to be 2.3 and 3.7, respectively. Stoichiometrically, benzene formation from benzenethiol neither requires nor deposits extra hydrogen. Hence the total integrated H_2 in the benzenethiol temperature programmed reaction spectrum is proportional to that benzenethiol which ultimately decomposes to surface atomic carbon via β_2-H_2 evolution. The stoichiometry of this intermediate decomposing to β_2-H_2 may be calculated to be $C_6H_{3.7}$, nearly the same as the stoichiometry of surface benzyne, C_6H_4, but not that of a surface phenyl group, C_6H_5.

Geometric and thermodynamic arguments further support the formation of surface benzyne. The intermediate which decomposes to β_2-H_2 is unusually stable. The

Figure 10. Model of the phenyl thiolate intermediate at saturation coverage. The sulfur is coordinated to the pseudo-threefold site with a Mo-S bond distance of 2.44 Å based on analogy with organometallic cluster analogues.

temperature of evolution, 680 [K], indicates that the activation energy for β_2-H_2 formation is approximately 41 [kcal/mol], greater by 6 [kcal/mol] than the activation energy for any other measured hydrocarbon fragment decomposition process on Mo(110). Any C-H bond activation step almost certainly requires close (approximately bonding distance) approach of the C-H bond to the surface somewhere along the reaction coordinate.[31] Hence, it is reasonable to assert that the intermediate decomposing to β_2-H_2 is constrained by its geometry to keep its C-H bonds inaccessible to the surface. By analogy to organometallic complexes, both surface benzyne[32,33,34] and the chemisorbed phenyl group[35] would be expected to bond with the aromatic ring oriented very nearly perpendicular to the surface. All C-H bonds are located relatively far from the surface in perpendicularly oriented surface benzyne (Fig. 6). If a surface phenyl group were present and oriented perpendicular to Mo(110), then two of the C-H bonds would be directed toward the surface. Therefore, of the two intermediates considered here, only surface benzyne would be expected to show the unusual stability that is experimentally observed.

Although benzyne is an unprecedented surface species, numerous examples of organometallic benzyne complexes have been reported. In these complexes, which range from mononuclear to trinuclear, benzyne bonds with the aromatic ring approximately perpendicular to the metal atom(s). NEXAFS spectra, shown in Fig. 11, are consistent with perpendicular benzyne azimuthally oriented along the (110) direction. The azimuthal dependence of the NEXAFS data must be measured to confirm the structure of the benzyne, however. In organometallic complexes, benzyne acts both as an η^2 donor via the formal carbon-carbon triple bond, and as an electron acceptor into unfilled π^* molecular orbitals. As a result of electron acceptance, extensive rehybridization of the benzyne carbon-carbon triple bond is inferred to occur. Hence, the best valence bond representation of these complexes invokes a di-σ bonded benzyne intermediate. Benzyne is envisioned to bond to Mo(110) similarly, as is depicted in Figure 6.

The reaction scheme deduced for benzenethiol is a general one for alkyl thiols on Mo(110). Surface alkyl thiolates have been identified both on Mo(110)[36] and on other surfaces[37]. Therefore, the formation of a strongly bound thiolate is anticipated on most transition metal surfaces with the exception of the some of the clean Group IB metals (Ag, Au) This strongly bound thiolate is stable with respect to desorption but not necessarily with

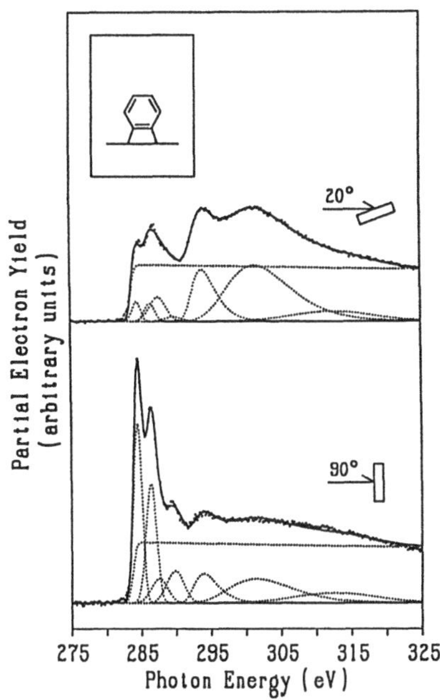

Figure 11. NEXAFS data for benzyne formed from phenyl thiolate on Mo(110). The data was collected subsequent to annealing benzenethiol multilayers to 600 [K].

respect to decomposition. Both are important considerations in lubrication and adhesion. On Mo(110), alkyl thiolates react via C-S hydrogenolysis at approximately 320 [K] to form gaseous hydrocarbons (alkanes and alkenes) and surface sulfur. Importantly, the stability of the thiolates with respect to C-S bond hydrogenolysis is independent of alkyl chain length for the primary thiols[38]. There is, however, a notable difference in the stability of the phenyl thiolate and t-butyl thiolate[39] compared to the primary alkyl thiols. The t-butyl thiolate reacts most rapidly and the phenyl thiolate most slowly. This pattern is consistent with a transition state for hydrogenolysis which has a considerable amount of C-S bond breaking character since the relative rates correlate with radical stability[40]. Therefore, the most stable thiolate moieties will be those with phenyl rings bound directly to the S.

6. Phenol Chemistry at Saturation Coverage

The reactivity and structure of phenol and intermediates derived from its reaction on Mo(110) are interesting to compare to the sulfur analogue, benzenethiol. Phenol is also a good model for metal-adhesive bonding since phenolics are a component in adhesives.[41] Therefore, the bonding and stability of phenol and intermediates formed from its reaction on the surface are of importance in a microscopic picture of adhesion. A comparison of phenol bonding and reactivity to benzenethiol also yields information about what reactive steps control adsorbate stability in this class of molecules. The C-O bond strength in phenol (110 [kcal/mol]) is 23 [kcal/mol] stronger than the corresponding C-S bond in benzenethiol (87 [kcal/mol])[40]. Therefore, if C-X (X=O,S) bond breaking determines the rate of reaction on the surface, the stability of O-containing intermediates should be greater than the corresponding S-containing species. Indeed, the stability of the phenoxide intermediate formed from O-H bond dissociation in adsorbed phenol is substantially more stable at high coverage than the sulfur analogue, phenyl thiolate[42].

Phenol may interact with metal surfaces by bonding through an oxygen lone pair as an electron donor or by interaction of the π-ring system with the surface analogous to the phenyl thiolate. Clearly, the type of bonding interaction will lead to different adsorption structures and reactivity. In addition to this work on Mo(110), d_0- and d_6-phenol adsorbed from aqueous solution have been investigated on Pt(111)[43]. These studies suggest that the O-H(O-D) bond is broken upon adsorption and that a phenoxide surface species is adsorbed parallel to the surface. Preliminary results from the reaction of phenol on Rh(111) show that the O-H bond cleavage occurs below 300 [K] to yield phenoxide. However, the phenoxide stability is considerably greater than on Mo(110) and, in fact, the C-O bond does not break on this surface, instead reaction to CO, H_2 and surface carbon is observed[27].

Phenol undergoes molecular desorption and irreversible decomposition on Mo(110), producing only gaseous H_2 and phenol. Carbon and oxygen remain on the surface after decomposition. No gaseous hydrocarbons are produced during temperature programmed reaction of phenol on Mo(110) as shown in Fig. 12. The temperature programmed reaction spectrum of the parent ion, 94 amu, exhibits two molecular desorptions at temperatures of 210 [K] and 240 [K]. The lower temperature peak at 210 [K] is attributed to multilayer desorption since this peak increases in intensity indefinitely with exposure. The desorption at 240 [K] does not exchange with preadsorbed surface deuterium atoms or influence the production of dihydrogen, thus it is assigned to desorption of a weakly bound molecular layer. Temperature programmed reaction of d_6-phenol and d_5-phenol, (C_6D_5OH), produced molecular desorption spectra identical to those obtained for h_6-phenol and yielded only those isotopes of phenol present in the dose.

The saturation coverage of irreversibly bound phenol is $\sim$1/9 of a monolayer, as measured by the peak-to-peak height of the O(KLL) Auger signal after reaction of a saturation exposure of phenol to 850 [K] where only atomic oxygen and carbon remain on the surface. Since no oxygen-containing products are desorbed, the amount of atomic oxygen remaining is a measure of the saturation coverage of irreversibly bound phenol. An ordered oxygen layer with a coverage of 1/3 was used as a reference state. Therefore, the saturation coverage of irreversibly bound phenol is similar to that of the phenyl thiolate.

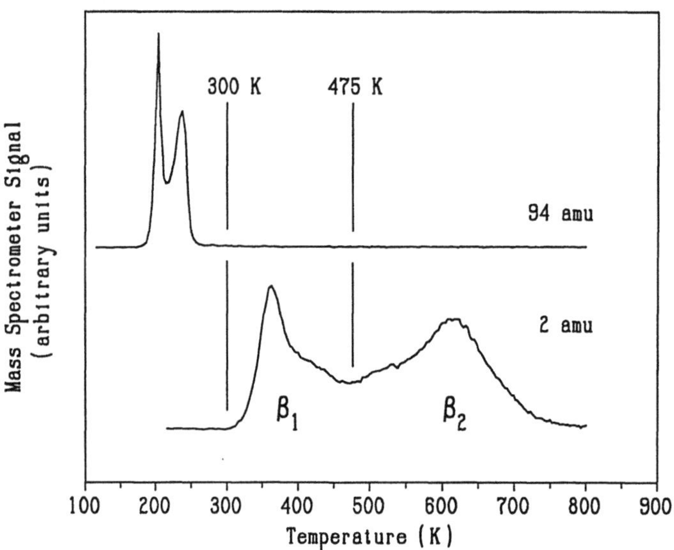

Figure 12. Temperature programmed reaction data for phenol on Mo(110) at high coverage.

No low energy electron diffraction patterns except that of Mo(110) were observed after reacting a saturation dose of phenol to 300, 500 or 850 [K], suggesting that overlayers with long-range order do not form for these surface species under the conditions of these experiments.

In this work, we use selective isotopic labelling to demonstrate that O-H bond cleavage occurs prior to C-H(D) bond cleavage. Temperature programmed reaction data for H_2, HD and D_2 formation was collected following saturation of Mo(110) with d_5-phenol at a crystal temperature of 100 [K]. H_2 is formed exclusively in the lowest temperature peak at 360 [K]. C-D bond cleavage must occur around 370 [K] since HD and D_2 begin to desorb at this temperature. Furthermore, the cleavage of more than one C-D bond below 475 [K] is evidenced by the D_2 evolved in a broad peak at 430 [K].

Temperature programmed reaction of a $\sim$1:1 mixture of d_6-phenol and h_6-phenol further demonstrates that O-H and O-D bonds are cleaved below 360 [K]. Since evolution of all isotopes of dihydrogen commences at the same temperature (360 [K]) in this experiment, O-H(D) bond cleavage is *not* the rate limiting process in the initial production of dihydrogen. Since we know that O-H bonds are broken first from the d_5-phenol reactions, the absence of an isotope effect is proof that O-H(D) bonds are broken below the temperature required for β_1-dihydorgen formation. Recall that in the analogous experiment with C_6H_6 and C_6D_6, an isotope effect in the β_1-dihydrogen formation and molecular benzene desorption kinetics was observed[8].

X-ray photoelectron spectroscopy demonstrates that although O-H bonds are broken below 360 [K] C-O bonds remain intact. The C(1s) data, shown in Fig. 13, demonstrates that there are intact C-O bonds at surface temperatures of 300 [K] based on the observation of a C(1s) peak with a binding energy of 285.6 [eV]. The C(1s) binding energy shift measured for the carbons bound directly to hydroxyl groups in the compound 1,3,5-trihydroxybenzene is 1.8 [eV] with respect to the carbons not bound directly to a hydroxyl group,[44] consistent with the shift observed here of 1.5 [eV] between the high binding energy carbon, 285.6 [eV], and the low binding energy carbon at 284.1 [eV]. The C(1s) binding energy of methoxy groups on Pd(111) has been measured to be 285.9 [eV],[45] in close agreement with the binding energy of 285.6 [eV] observed here for carbons bound to oxygen. Thus the higher binding energy carbon can clearly be associated with the carbon bound to oxygen in phenol.

More than one oxygen containing species is present on the surface at 300 [K], as shown by the high coverage oxygen (1s) X-ray photoelectron data. Two oxygen (1s) peaks are detected at 532.3 [eV] and 530.8 [eV] after heating phenol multilayers to 300 [K]. The two O(1s) peaks vary in relative intensity as a function of coverage and surface temperature, demonstrating that the two different peaks correspond to chemically inequivalent species and are not due to final state effects. Temperature programmed reaction data shows that C-H bond activation does not occur until 370 [K], ruling out surface species which have undergone C-H bond cleavage such as a-dehyro-phenoxide which would be bound to the surface via an a carbon and the oxygen. The binding energies of alkoxide species on Cu(110) formed from primary alcohols have been measured to be in the range of 530.7 to 531.0 [eV][46]. Thus the binding energy of 530.8 [eV] is in the range expected for a surface phenoxide species. The binding energy of 532.3 [eV] is slightly above the range of binding energies measured for a series of alcohols on Cu(110): 532.7 to 533.0 [eV]. Given that there are two oxygen-containing species on the surface, and that these species produce a C(1s) spectrum similar to that of phenol multilayers, the two species most likely to be present are phenoxide (C_6H_5O-) and either phenoxide adsorbed in a different site or molecular phenol. Thus either phenol or phenoxide bound at a different site may account for the second O(1s) peak. This issue could be resolved by vibrational studies which have not yet been performed, and again, demonstrates the importance of using complementary techniques.

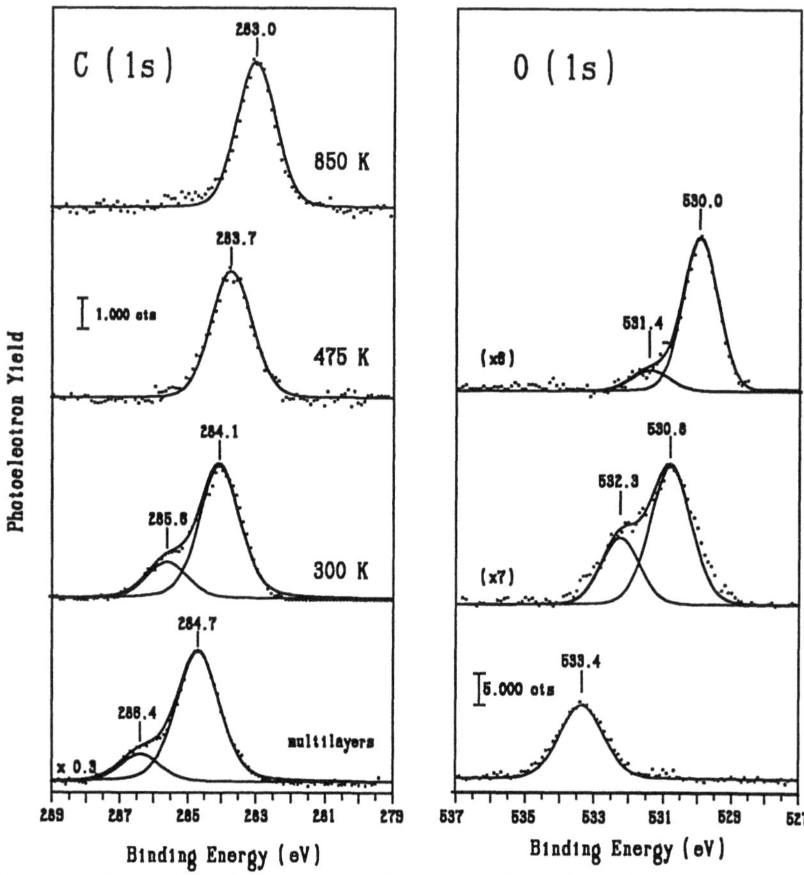

Figure 13. X-ray photoelectron spectra for intermediates derived from phenol on Mo(110). The data were obtained by annealing phenol multilayers to the temperatures indicated. Both C(1s) and O(1s) data were collected in the same experiment.

Carbon-oxygen bond cleavage commences in the range of 360-400 [K], indicating an enhanced stability of the phenoxide compared to phenyl thiolate which began to cleave C-S bonds at 250 [K] at saturation coverage. Atomic oxygen is first observed in the X-ray photoelectron data after annealing to temperatures between 350 and 375 [K]. After annealing to 475 [K], only atomic oxygen is observed in the X-ray photoelectron spectra. The C(1s) X-ray photoelectron data further indicate that carbon-oxygen bonds are cleaved by 475 [K]. Only one C(1s) peak with a binding energy of 283.7 [eV] with a full width at half maximum of 1.4 [eV], characteristic of hydrocarbon species with intact C-C and C-H bonds[8], is observed after annealing to 475 [K], the temperature between the β_1- and β_2-H$_2$ peaks. Clearly, C-O bond cleavage has occurred by this temperature since the high binding energy carbon is absent and the O(1s) spectrum is that of atomic oxygen. We further note that the data is *not* consistent with the presence of surface benzyne since no asymmetry in the C(1s) peak is observed and the C(1s) binding energies are not the same as for benzyne. Thus, both the stability of the X-(C$_6$H$_5$) intermediates and the product distribution for their reaction are dramatically altered by substitution of the heteroatom.

Total decomposition to atomic constituents occurs by 850 [K]. Only atomic carbon is detected at its characteristic binding energy of 283.0 [eV] in the C(1s) spectrum obtained after annealing to 850 [K], a temperature past that of all dihydrogen evolution. Recall, that only atomic oxygen is observed in the O(1s) data after annealing to 475 [K] or above.

The observed difference in the reactivity of phenoxide and phenyl thiolate correlates with the relative C-O and C-S bond strengths. The difference in reactivity is not attributable to a difference in the heat of adsorption of atomic oxygen and sulfur on Mo(110), which could influence the relative kinetic stability of the phenoxide and phenyl thiolate surface species. The heat of adsorption of sulfur on Mo(110) is -105 [kcal/mol] [47]. and although the heat of adsorption of oxygen has not been measured on Mo(110), we expect it to be equal to or more negative than that of sulfur. We base this assumption on data from several sources. First, on Pt(111), one of the few surfaces for which the heat of adsorption of both oxygen and sulfur have been experimentally measured, the heat of adsorption of oxygen is slightly more negative than that of sulfur, -85 [kcal/mol] for oxygen[48] and -80 [kcal/mol] for sulfur[47]. Second, comparison of the heat of formation of MoO_3 and MoS_2, -180 [kcal/mol] and -56 [kcal/mol], respectively, suggests that Mo-O bonds are on the average stronger than Mo-S bonds after correction for the coordination number in each compound yielding average heats of formation of -60 [kcal/mol] for MoO_3 and -28 [kcal/mol]e for MoS_2. A correlation between the corrected heats of formation of metal carbides, oxides and nitrides and heats of adsorption of atomic species has previously been proposed and found to correctly predict dissociative adsorption of diatomic molecules.[49] Finally, the strength of the metal-sulfur and metal-oxygen bonds in the organometallic analogues, $Ti(Cp)_2(SC_6H_5)_2$ and $Ti(Cp)_2(OC_6H_5)_2$, have been calculated to be 86 [kcal/mol] and 111 [kcal/mol], respectively[50]. All of the above suggest that the heat of adsorption of oxygen should be more negative than that of sulfur on Mo(110). Thus the relative heat of adsorption of oxygen and sulfur cannot account for the observed difference in reactivity since phenoxide would be predicted to be *more* stable with respect to carbon-heteroatom bond cleavage than phenyl thiolate; this is the opposite of the observed relative stabilities of the two species.

The difference in reaction selectivity observed for phenyl thiolate and phenoxide is attributed to the cleavage of C-H bonds in the same temperature regime as C-O bond cleavage in phenoxide. In both phenoxide and phenyl thiolate, C-H(D) bond activation occurs near 370 [K] as evidenced by the observation of benzyne by X-ray photoelectron spectroscopy in the benzenethiol studies and by the evolution of D_2 from reaction of d_5-phenol at 370 [K] in the temperature programmed reaction spectrum from the reaction of phenol. The greater carbon-heteroatom bond strength causes phenoxide to undergo C-O bond cleavage in the same temperature regime as C-H bond activation and the concomitant formation of gaseous dihydrogen, resulting in decomposition of the intermediate rather than reaction to benzene and surface benzyne. Furthermore, the surface is depleted of surface hydrogen via hydrogen atom recombination in the temperature range where C-O bond cleavage commences making hydrogenolysis unlikely. Since phenyl thiolate begins to undergo C-S bond cleavage at 250 [K], a temperature almost 100 [K] lower than the onset of C-H bond activation and where surface hydrogen from scission of the S-H bond is still present on the surface in significant amounts, the thiolate species reacts to form benzene.

Structural effects may also play a role in determining differences in the reactivity of phenoxide and phenyl thiolate. Unfortunately, the structure of the phenoxide intermediate is not amenable to determination by NEXAFS or other methods since a mixture of species is always present. We note that had the X-ray photoelectron experiments not been performed and a single species assumed, an erroneous structure with contributions from both surface intermediates would be obtained which would not be representative of a single phenoxide. This example demonstrates the limitations of surface structural determinations for complex adsorbate systems that populate more than one state.

Although, the adsorption structure of the phenoxide intermediate is unknown and is not directly amenable to determination with the NEXAFS method, qualitative information can be obtained from a consideration of the coverage of phenoxide. The saturation coverage of irreversibly bound phenol is comparable to but smaller than the phenyl thiolate but larger than that of parallel bound benzene. The relative saturation coverage of benzene, phenoxide and phenyl thiolate are 1.0: 1.4 :1.6. Therefore, all of the irreversibly bound phenol can not be bound parallel to the surface plane, on steric grounds. It is possible, however, that a mixture of parallel and perpendicular species may be present accounting for the two inequivalent species observed in X-ray photoelectron spectroscopy. A parallel bonding geometry has been suggested on Pt(111) based on the coverage of phenol at saturation[43]. We note that a parallel bound phenoxide is not expected to form benzyne with high selectivity since the nearly perpendicular geometry of the phenyl thiolate is deemed important in selective ortho dehydrogenation to form benzyne.

7. <u>Coverage</u> <u>Dependence</u> <u>of</u> <u>Adsorbate</u> <u>Stability</u>

The reaction kinetics of both phenyl thiolate and phenoxide on Mo(110) are strongly dependent on coverage. Notably, there is a self-stabilization of these intermediates at high coverage which qualitatively alters the reaction selectivity. We emphasize that the presence of sulfur or oxygen overlayers is not necessary for the stabilization of these intermediates, although surface modification also affects the kinetics for reaction. In both cases, the rate of decomposition is more rapid at low coverage. In the case of the phenyl thiolate, C-S, C-C and C-H bond cleavage are more rapid at low coverage. In the case of phenoxide, the rates of C-C and C-H bond cleavage are faster at low coverage but the C-O bond activation kinetics are essentially unchanged.

The coverage dependence in the stability of the phenyl thiolate on Mo(110) dramatically effects the reaction selectivity. At high coverage, as discussed above, selective formation of benzene and benzyne occur, consistent with the nearly perpendicular ring orientation. Temperature programmed reaction data obtained as a function of phenyl thiolate coverage is shown in Fig. 14. At low coverage, nonselective decomposition to H_2 and surface carbon and sulfur predominates. X-ray photoelectron data demonstrates that nonselective decomposition is accompanied by C-S bond cleavage at low temperature with all C-S bonds cleaved by 225 [K] at the lowest coverage investigated (Fig. 14). Recall that at saturation coverage, C-S bond scission commenced at 250 [K] and was not complete until 550 [K].

NEXAFS data obtained at intermediate coverages suggests the presence of two types of phenyl thiolate intermediates, one parallel bound and one nearly perpendicular to the surface plane. This suggests that the coverage dependence in the reaction selectivity and kinetics may be related to a coverage dependent structural transformation. A parallel ring orientation is expected to give rise to nonselective decomposition, analogous to benzene, whereas a more upright geometry is expected to react selectively to produce benzene and benzyne. NEXAFS experiments must be performed at very low coverage in an attempt to isolate the parallel bound state to determine if this model is correct. The suggestion that intermolecular interactions may determine the stability of organic molecules is of potential importance in adhesion and lubrication.

Interestingly, the temperature of C-O bond cleavage in phenoxide is not changed at lower coverage as determined by X-ray photoelectron spectroscopy. Instead, C-C bond cleavage occurs at lower temperature for lower coverages. Low coverage (0.45 saturation) X-ray photoelectron data shows that surface species similar to those observed at high coverage are present at low temperature and low coverage, however, C-C bond cleavage occurs at lower temperature than C-O bond cleavage. The temperature required for C-O bond cleavage is approximately the same temperature (T < 475 [K]) at high and low coverage. X-ray photoelectron data were collected at an exposure 0.45 of saturation. The C(1s) spectrum

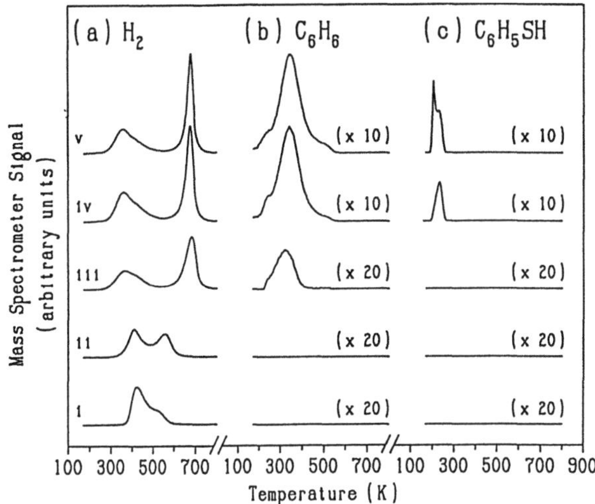

Figure 14. Temperature programmed reaction of benzenethiol on Mo(110) as a function of coverage.

obtained at 100 [K] is similar to that observed at 300 [K] for saturation doses (Fig. 13). After annealing to 300 [K], carbon with a binding energy of 283.0 [eV] appears in the X-ray photoelectron spectra, indicating that some C-C bond cleavage has occurred. *Some* higher binding energy carbon (285.5 [eV]), characteristic of carbon bound to oxygen, is still present at this temperature. The O(1s) data indicate that at 100 [K], most of the O(1s) intensity appears at 530.8 [eV] consistent with the presence of phenoxide. Atomic oxygen (530.0 [eV]) does not appear until after annealing to 400 [K], consistent with the observed C(1s) spectra. There is also a coverage dependence to the H$_2$ formation during temperature programmed reaction, consistent with more facile C-H bond breaking at low coverage.

8. Conclusions

The reactivity and structure of benzene, phenyl thiolate and phenol on Mo(110) have been compared. Benzene is bound parallel to the surface plane and primarily undergoes nonselective decomposition to H$_2$ and surface carbide. The phenyl thiolate is bound with the ring more nearly perpendicular to the surface plane at high coverage. The nearly perpendicular ring orientation is proposed to be important in the selectivity for hydrogenolysis and ortho dehydrogenation. The structure of the phenoxide intermediate was not measured, but the stability of the phenoxide is greater than the phenyl thiolate, consistent with the relative values for the C-S and C-O bonds. This result suggests that C-X (X=S,O) bond breaking controls reaction selectivity and kinetics for this class of molecules on Mo.

The kinetics for C-S and C-H bond scission in the phenyl thiolate and C-C and C-H bond cleavage in phenoxide are strongly dependent on the coverage of these intermediates. In both cases, decomposition proceeds more rapidly at low coverage. In the case of the phenyl thiolate, a coverage dependent structural transformation is suggested to be important in the coverage dependence of its stability. These results point out the importance of intermolecular interactions in determining the stability of a metal-adhesive interface.

9. <u>Acknowledgements</u>

The able assistance of A.C. Liu, J.G. Serafin, J.T. Roberts, and J. Stohr is acknowledged in the performance of these experiments. ACL and JGS are also thanked for their comments and discussion of this manuscript. This work was supported by the Department of Energy, Basic Sciences Division, Grant No. DE-FG02-84ER13289.

10. <u>References</u>

1. M.A. Van Hove and S.Y. Tong, <u>Surface Crystallography by LEED</u> (Springer, Heidelberg, 1979).

2. N.V. Richardson and A.M. Bradshaw, <u>Electron Spectroscopy</u> C.R. Brundle and A.D. Baker, eds. (Academic, NY (1981)) Vol. 5, Chapter 3.

3. J.Stohr and D. Outka, Phys. Rev. B 36 (1987) 7891.

4. J.Stohr and R. Jaeger, Phys. Rev. B 26 (1982) 4111.

5. a.) J. Stohr, E.B. Kollin, D.A. Fischer, J.B. Hastings, F. Zaera, F. Sette, Phys. Rev. Let. 55 (1985) 1468; b.) J. Stohr, J.L. Gland, E.B. Kollin, R.J. Koestner, A.L. Johnson, E.L. Muetterties, F. Sette, Phys. Rev. Lett. 53 (1984) 2161.

6. a.) A.L. Johnson, E.L. Muetterties, J. Stohr, J. Phys. Chem. 89 (1985) 2161; b.) M. Bader, J. Haase, K.H. Frank, A. Puschmann, A. Otto, Phys. Rev. Lett. 56 (1986) 1921.

7. A.L. Johnson, E.L. Muetterties, J. Stohr, J. Am. Chem. Soc. 105 (1983) 7183.

8. A.C. Liu and C.M. Friend, J. Chem. Phys. 89 (1988) 4396.

9. R.J. Koestner, J. Stohr, J.L. Gland, E.B. Kollin, F. Sette, Chem. Phys. Lett. 120 (1985) 285.

10. J.Stohr, J.L. Gland, W. Eberhardt, D. Outka, R.J. Madix, F. Sette, R.J. Koestner, U. Doebler, Phys. Rev. Lett., 51 (1983) 2414.

11. A.C. Liu, J. Stohr, C.M. Friend, R.J. Madix, In preparation.

12. D. Outka and J.Stohr, J.Chem. Phys. 88 (1988) 3539.

13. a.) L.L. Kesmodel, R.C. Baetzhold, and G.A. Somorjai, Surf. Sci., 66 (1977) 299; b.) L.L. Kesmodel, L.H. Dubois, J. Chem. Phys. 70 (1979) 2180.

14. R.J. Madix, Cat. Rev. 26 (1984) 281 and references therein.

15. W.L. Jolly <u>Electron Spectroscopy</u>, C.R. Brundle and A.O. Baker, (Academic Press, NY (1978)) p 119.

16. H. Ibach and D.L. Mills, <u>Electron Energy Loss Spectroscopy and Surface Vibrations</u> (Academic Press, NY, 1982).

17. a.) C.M. Friend and E.L. Muetterties, J. Am. Chem. Soc. 103 (1981) 773; b.) A.K. Myers, G.R. Schoofs, J.B. Benziger, J. Phys. Chem., 91 (1987) 2230; c.) P.M. Blass, S. Akhter, J.M. White, Surf. Sci. 191 (1987) 406.

18. a.) G.D. Waddill and L.L. Kesmodel, Phys. Rev. B, 31 (1985) 4940; b.) V.H. Grassian and E.L. Muetterties, J. Phys. Chem., 91 (1987) 389.

19. F. Netzer, Phys. Rev. B, 37 (1988) 10399.

20. a.) M.-C. Tsai and E.L. Muetterties, J. Am. Chem. Soc. 104 (1982) 2534; b.) M.-C. Tsai and E.L. Muetterties, J.Phys. Chem. 86 (1982) 5067; c.) S. Lehwald, H. Ibach, J.E. Demuth, Surf. Sci., 78 (1978) 577.

21. M. Bader, J. Haase, K.-H. Frank, C. Ocal, A. Puschmann, J. Phys. (Paris) 47 (1986) C8-491.

22. J.T. Roberts, R.J. Madix, In preparation.

23. a.) M.A. Van Hove, R.F. Lin and G.A. Somorjai, Phys. Rev. Lett. 51 (1983) 778; b.) B.E. Koel, J.E. Crowell, C.M. Mate, G.A. Somorjai, J. Phys. Chem. 88 (1984) 1988.

24. J.A. Polta, P.A. Thiel, J. Am. Chem. Soc. 108 (1986) 7560.

25. J.P. Collman, L.S. Hegedus, J.R. Norton, R.G. Finke, <u>Principles</u> <u>and</u> <u>Applications</u> <u>of</u> <u>Organotransition</u> <u>Metal</u> <u>Chemistry</u> (University Science Books, Mill Valley, CA (1987).

26. A.K. Meyers, G.R. Schoofs, and J.B. Benziger, J. Phys. Chem. 91 (1987) 2230.

27. X. Xu and C.M. Friend, In preparation.

28. J.T. Roberts and C.M. Friend, J. Chem. Phys. 88 (1988) 7172.

29. R.G. Nuzzo, B.R. Zergaski, L.H. Dubois, J. Am. Chem. Soc. 109 (1987) 733 and references therein.

30. J.T. Roberts, P.A. Stevens and R.J. Madix, In preparation.

31. E.L. Muetterties, R.M. Wexler, Surv. Prog. Chem., 10,(1983), 61.

32. S.L McLain, R.R. Schrock, P.R. Sharp, M.R. Churchill, W. Y. Youngs, J. Am. Chem. Soc. 101 (1979), 263.

33. M. D. Rousch, R.G. Gastinger, S.A. Gardner, R.K. Brown, J.S. Wood, J. Am. Chem. Soc. 99 (1977) 7870.

34. R.G. Goudsmit, B.F.G. Johnson, J. Lewis, P. Raithby, M.J. Rosales, J. Chem. Soc. Dalton Trans.(1983), 2257.

35. R.W. M.ten Hoedt, J.G. Noltes, G. van Koten, A.L. Spek, J. Chem. Soc. Dalton Trans.(1978), 1800.

36. a.) J.T. Roberts and C.M. Friend, J. Phys. Chem. 92 (1988) 5205; J.T. Roberts and C.M. Friend, Surf. Sci., 198, (1988) L321.

37. a.) R.J. Koestner, J.Stohr, J.L. Gland, E.B. Kollin, F. Sette, Chem. Phys. Lett., 120 (1985) 285; b.) B.A. Sexton, G.L. Nyberg, Surf. Sci. 165 (1986) 251; c.) T. Edmonds, J.J. McCarroll, R.C. Pitkethyly, Ned. Tijdschr. Vacuumtech. 8 (1970) 162; d.) D.R. Huntley, In preparation.

38. J.T. Roberts and C.M. Friend, J. Am. Chem. Soc. 109 (1987) 4423.

39. B.A. Wiegand, C.M. Friend, J.T. Roberts, In preparation.

40. S.W. Benson, Chem. Rev. 78 (1978), 23.

41. J. Robins, in: Structural Adhesives: Chemistry and Technology (Plenum, New York, 1986).

42. J.G. Serafin, C.M. Friend, Surf. Sci., In press, (1988).

43. F. Lu, G.N. Salaita, L. Laguren-Davidson, D.A. Stern, E. Wellner, D.G. Frank, N. Batina, D.C. Zapien, N. Walton and A.T. Hubbard, Lang. 4 (1988) 637.

44. U. Gelius, P.F. Heden, J. Hedman, B.J. Lindberg, R. Manne, R. Nordberg, C. Nordling and K. Siegbahn, Physica Scripta, 2 (1970) 70.

45. R.J. Levis, J. Zhicheng and N. Winograd, J. Am. Chem. Soc. 110 (1988) 4431.

46. a. M. Bowker and R.J. Madix, Surf. Sci. 95 (1980) 190. b. M. Bowker and R.J. Madix, Surf. Sci. 116 (1982) 549.

47. J. Benard, J. Oudar, N. Barbouth, E. Margot and Y. Berthier, Surf. Sci. 88 (1979) L35.

48. C.T. Campbell, G. Ertl, H. Kuppers and J. Segner, Surf. Sci. 107 (1981) 220.

49. J.B. Benziger, Appl. Surf. Sci. 6 (1980) 105.

50. M. Calhorda, M. de C.T. Carrondo, A. Dias, A. Domingos, J. Simones and C. Teixeira, Organomet. 5 (1986) 660.

Surface Chemistry of Perfluoropolyethers and Hydrogenated Analogs: Are Studies of Model Compounds Useful?

M.M. Walczak and P.A. Thiel

Department of Chemistry and Ames Laboratory
Iowa State University, Ames, IO 50011, USA

We have studied adsorption, desorption, and decomposition of ethers on
Ru(001), an atomically-smooth metal surface. We have compared diethers
with monoethers, and fluorinated ethers with hydrogenated ethers. The
number of ether linkages does not strongly influence adsorption bond
strength, nor the extent of decomposition. Fluorination does weaken the
adsorption bond strength and prevents decomposition. These studies suggest
that the surface properties of monomeric ethers can be used to predict
properties of oligomeric, and perhaps even polymeric, ethers.

1. Introduction

Polymeric, fluorinated ethers are marketed as industrial lubricants, under
trade names such as Krytox [1], Demnum [2], and Fomblin [3]. In some
applications, the surface chemistry of the ether is quite important to its
proper function as a lubricant. An example of such an application is in
the lubrication of computer disks, where thin layers (perhaps only one
molecule thick) of lubricant serve to protect the disk from the head. The
lubricant is often non-replenishable, or replenishable only to a limited
extent, so that its loss by any mechanism can be catastrophic. The
lubricant must therefore adhere strongly to the substrate, to avoid being
swept off by centrifugal force. The lubricant must also resist chemical
degradation. In this application, then, the important issues of surface
chemistry are the adsorption bond strengths and decomposition reactions of
the ethers.

Another area in which these compounds find application is the
aeronautical and space industry, where several physical properties combine
to make them favored choices as bearing lubricants [e.g. 4]. Here, too,
surface chemistry is important, since the perfluoropolyethers can undergo
decomposition reactions which are (apparently) catalyzed by metal surfaces
[5-7, and references therein]. The reactions are accompanied by evolution
of gases and corrosion of the metal. The important issue for surface
science in this case is identification of the decomposition mechanism, and
identification of those factors which stabilize the ether against catalytic
decomposition.

In principle, surface science is ideally poised to address issues such
as these, via studies of adsorption bond strengths and decomposition
pathways in model systems. As yet, however, the techniques of modern
surface science have been rarely applied to fundamental studies of
fluorinated ethers. In fact, there have been only two reports in the
literature. In one of these, we studied perfluorodiethyl and
perhydrodiethyl ether on Ru(001) [8]. We found that fluorination weakens
the chemisorption bond to the metal, presumably because fluorination
retards electron donation into the metal from the oxygen lone pairs of the
ether [8]. This result parallels that of Avery [9], who studied

Springer Series in Surface Sciences, Vol. 17
Adhesion and Friction   Editors: M. Grunze and H.J. Kreuzer
© Springer-Verlag Berlin, Heidelberg 1989

hexafluoroacetone (a ketone which can bond to the surface in a fashion
similar to an ether, i.e. via electron donation from a lone pair). This
molecule bonds more weakly to Pt(111) than does the hydrogenated analog
[9]. Another study was that of Ng et al. [10], which focused on dimethyl
ether and its partially-fluorinated analog, $(CF_2H)_2O$, on Al_2O_3. They found
that neither species undergoes detectable decomposition. The desorption
temperature of the fluorinated ether indicates an adsorption bond strength
of 33 kJ/mol. Their results, for a partially-fluorinated ether adsorbed on
a metal oxide substrate, Al_2O_3, are roughly comparable to our own results
for a perfluorinated ether adsorbed on the metallic Ru(001) substrate.
This may indicate that the chemistry of ethers on metal oxides will not
differ grossly from that of ethers on metals, although more extensive work
is certainly necessary before this conjecture is on firm ground.

The present paper describes an extension of our work, from simple
diethyl ether, to larger and more complex molecules. These larger
molecules, dimeric ethers, are one step closer to the commercial
polyethers. By comparing the surface chemistry of monomeric and dimeric
ethers, we begin to test the validity of using monomers as models for
polymeric compounds in these types of studies.

2. Experimental

The experiments are performed in an ion-pumped stainless steel UHV chamber.
The chamber base pressure is 7×10^{-11} Torr. The experimental apparatus and
methods are described in detail elsewhere [8,11]. The _T_hermal _D_esorption
Spectroscopy (TDS) technique is used to determine adsorption bond
strengths. The experiment consists of two steps. First, a surface is
dosed with the gas of interest. Second, the sample is heated in vacuum
while monitoring the gas phase. The temperatures at which the molecules
and/or decomposition products desorb from the surface are related to the
molecule-surface bond strength, and/or to the activation energy of the
decomposition reaction.

3. Results

Two sets of thermal desorption spectra are shown in Figs. 1 and 2. These
spectra are _representative_ of our results to date. These spectra are
obtained following adsorption of an ether on the atomically-smooth Ru(001)
surface. Figure 1 illustrates desorption of $CH_3CH_2OCH_2CH_3$, which we refer
to by its common name of diethyl ether. Figure 2 shows desorption spectra
of $CF_3CF_2OCF_2CF_3$, which we similarly call perfluorodiethyl ether. We have
also studied three other ethers on this surface: $CH_3CH_2OCH_2OCH_2CH_3$, which
we call diethoxymethane, $CH_3CH_2OCH_2CH_2OCH_2CH_3$, which we call diethoxyethane,
and $CF_3CF_2OCF_2CF_2OCF_2CF_3$, which we call perfluorodiethoxyethane. The ethers
which we have studied on Ru(001) are summarized in Table 1. Note that
diethoxymethane and diethoxyethane are _di_ethers, whereas diethyl ether is a
_mono_ether. The IUPAC-endorsed names of these compounds, and desorption
spectra of compounds other than diethyl ether, are available elsewhere
[11].

Thermal desorption spectra of a representative hydrogenated ether,
diethyl ether, are shown in Fig. 1. A single broad state, denoted α_1,
appears at ca. 200 K for low exposures of diethyl ether (curve a). For
intermediate diethyl ether exposures (curve b), a second state, α_2, at ca.
170 K is populated. At higher exposures a third state, γ, appears at lower
temperatures. This state is evident in curves c-d of Fig. 1. The γ-state
does not saturate with increasing exposure.

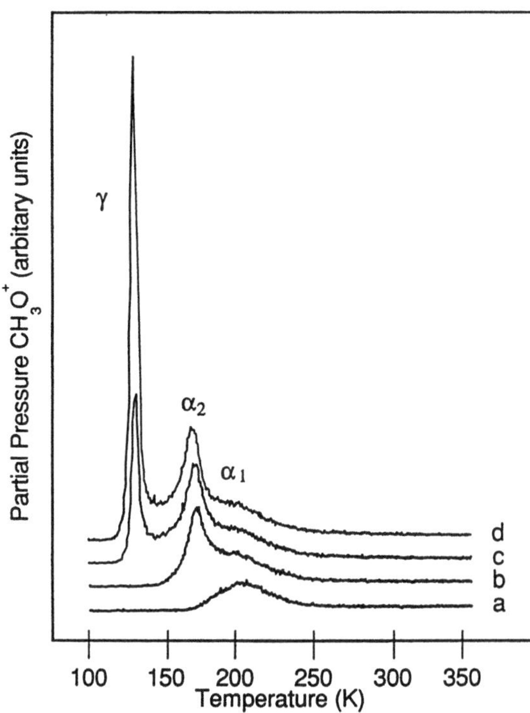

Figure 1. Thermal desorption spectra representative of hydrogenated ethers. Diethyl ether from Ru(001) following exposure of a) 0.13, b) 0.25, c) 0.38 and d) 0.50 L (1 L $\equiv$ 10^{-6} Torr*s).

The α-states exhibit typical first order desorption characteristics, including peak widths and temperatures which are approximately constant over the entire exposure range. We attribute the α-states to desorption of chemisorbed diethyl ether molecules. This conclusion is also based on the fact that the α-states saturate with increased exposure and have peak temperatures greater than that of the γ-state. Both α-states of diethyl ether on Ru(001) are quite broad (FWHM for α_1 = 54±13 K, α_2 = 20±3 K). This broadness is observed also in the chemisorption states of the other hydrogenated ethers [11].

The γ-state displays typical zero order desorption characteristics, including an increase in peak temperature and a decrease in peak width with increasing exposure. We attribute the γ-state to desorption from a condensed multilayer. This assignment is based on the inability to saturate the γ-state with increasing exposure, and its zero-order desorption characteristics. Multilayer desorption states are also observed for the other compounds described in Table 1.

Analysis of the α_1-state by Redhead's method for first-order desorption kinetics [12] yields a value of 51-53 kJ/mol for the <u>de</u>sorption barrier at low exposures of diethyl ether. There is no evidence that adsorption is appreciably activated, so we equate the desorption energy to the <u>ad</u>sorption bond strength. The α_2-state bond strength is 43-44 kJ/mol. The molecule-surface bond strengths for the other compounds listed in Table 1 are obtained similarly. Since the relative areas of the α_1- and α_2-peaks of

91

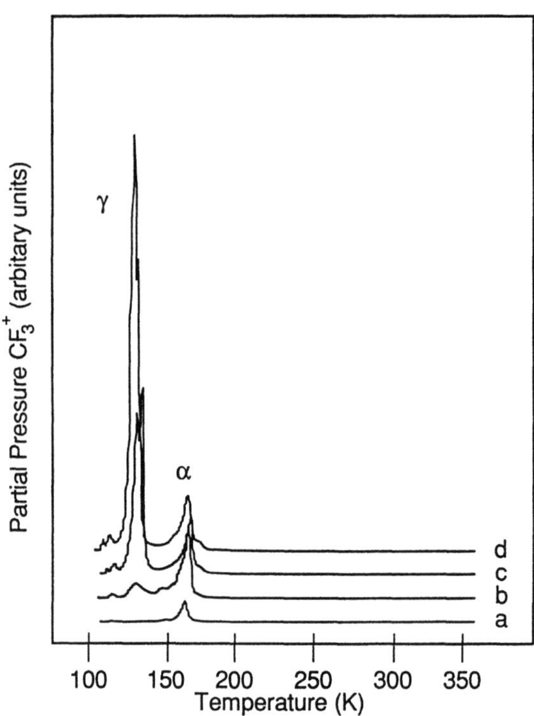

Figure 2. Thermal desorption spectra representative of fluorinated ethers. Perfluorodiethyl ether from Ru(001) following exposure of a) 0.07, b) 0.27, c) 0.67 and d) 1.0 L.

Table 1. Desorption Energies of Majority States: Hydrogenated vs. Fluorinated Ethers on Ru(001)

Compound	Hydrogenated (X=H) Energy (kJ/mol)	Fluorinated (X=F) Energy (kJ/mol)
Diethyl ether $CX_3CX_2OCX_2CX_3$	43-44;51-53	42-43
Diethoxymethane $CX_3CX_2OCX_2OCX_2CX_3$	53-69	-----
Diethoxyethane $CX_3CX_2OCX_2CX_2OCX_2CX_3$	58-62	44-47

diethyl ether indicate that about half the chemisorbed molecules occupy each state, both states are included in Table 2. Diethoxyethane also exhibits multiple chemisorbed states. The two minority states represent only 24% and 4% of the total desorption peak area. The desorption energies for the minority states, calculated as described above, are 73 and 90 kJ/mol, respectively. For purposes of comparison with other compounds, only the desorption energy of the majority state of diethoxyethane is used.

Ruthenium surfaces are known to catalyze the decomposition of oxygenated hydrocarbons [13-18]. We measure the extent of ether decomposition by measuring the amount of CO and H_2 which desorb from the surface [8,11]. Carbon monoxide and hydrogen coverages are calculated by comparing desorption peak areas with peak areas for saturation coverage. Saturation coverage for CO is 0.67 monolayers [19] and for hydrogen is 2 monolayers [20,21]. No desorption of other decomposition products is observed. We find that 0.04 to 0.17 monolayers of hydrogenated ethers decompose on the Ru(001) surface. These results are given in Table 2. Comparing the yields of O (as CO) and hydrogen for the <u>diethers</u> shows that oxygen must be quantitatively converted into CO in the decomposition reaction, since the ratio of CO to H agrees, within the experimental error, to that expected from the O:H stoichiometry in the parent compound. Therefore, the amount of ether which decomposes is half the CO yield for the <u>diethers</u>. When calculating the extent of decomposition for the monomeric diethyl ether, we find that the values obtained based on CO and hydrogen yields do not agree with the molecular stoichiometry. We use the higher of these two values (that based on hydrogen yield) for the extent of diethyl ether decomposition, 0.17 monolayers. Apparently, the oxygen in this molecule is not quantitatively converted to CO.

Table 2. Extent of Decomposition: Hydrogenated vs. Fluorinated Ethers on Ru(001)

Compound	Molecular CO Yield, Monolayers	Hydrogen Yield, Monolayers
Diethyl ether	0.04 ±0.02	1.7 ± 0.5
Diethoxymethane	0.09 ±0.02	0.5 ± 0.2
Diethoxyethane	0.07 ±0.02	0.7 ± 0.1
Perfluorodiethyl ether	<0.02	---
Perfluorodiethoxyethane	<0.02	---

Thermal desorption spectra of a representative fluorinated ether, perfluorodiethyl ether, are shown in Fig. 2. A sharp desorption feature, α, at ca. 165 K, is observed at low exposures (curves a-b). At higher exposures (curves c-d) another desorption feature, γ, emerges at ca. 130 K. The γ-state cannot be saturated with increasing exposure.

The α-peak exhibits first order desorption characteristics. It is attributed to chemisorbed perfluorodiethyl ether molecules for reasons analogous to those described above for diethyl ether. The desorption energy calculated for the narrow α-state (FWHM = 7 ± 1 K) is 42-43 kJ/mol (see Table 1). The γ-state displays typical zero order desorption kinetics and is attributed to desorption from a condensed multilayer of perfluorodiethyl ether.

In contrast to the hydrogenated ethers, the fluorinated ethers do not decompose significantly. The extent of decomposition, given in Table 2, is less than our detection limit for CO (0.02 monolayers).

4. Discussion

4.1 Fluorocarbons vs. Hydrocarbons

Previous investigations of ether-surface chemistry [22-24] indicate that ethers interact with metal surfaces in two ways. The stronger interaction consists of donation of electrons from the oxygen lone pair to the surface. This component contributes ca. 40 kJ/mol to the overall bond strength [22-25]. One would expect fluorination to weaken this ether-surface interaction due to the inductive withdrawal of electron density from the oxygen atom by the fluorinated alkyl groups.

The other component of the ether-surface bond is the alkyl-metal attraction. This interaction is much weaker than the oxygen-metal bond; each methylene group contributes 5 to 6.5 kJ/mol to the overall ether-surface bond [23,24,26,27]. This type of interaction is observed for alcohols [23,24] and cyclic hydrocarbons [26,27] as well as ethers [23,24]. Fluorination is expected to weaken this interaction as well, since the C-F bond is longer than the C-H bond [28], and fluorine is more electron rich than hydrogen [29]. The carbon is held further away from the surface by the first factor; fluorine-metal repulsion is important due to the second factor.

Our results confirm that, in the limit of low exposure, fluorination does weaken the ether-surface bond. As shown in Table 1, diethyl ether molecules bond ca. 10 kJ/mol more strongly than the fluorinated analog, perfluorodiethyl ether, on Ru(001). Furthermore, fluorination also weakens the ether-surface bonds of diethoxyethane on Ru(001). Therefore, the fact that fluorination weakens the ether-surface bond appears to be independent of the number of ether linkages.

The instability of chemisorbed hydrogenated ethers relative to fluorinated ethers is reflected in our data in three ways. First, the yield of decomposition products (CO and H_2) is measurable for hydrogenated ethers, whereas, the yield of CO for the fluorinated ethers is less than our detection limit (0.02 monolayers).

Second, the chemisorption peaks of the hydrogenated ethers are typically broad (c.f. Fig. 1), while the fluorinated ethers exhibit sharp peaks (c.f. Fig. 2). We suggest that the broadness of the peaks for hydrogenated ethers probably reflects the changing condition of the surface during desorption. In other words, desorption and decomposition are competing processes during the thermal desorption experiment.

The third trend is that the desorption yield of the hydrogenated ethers is relatively low in the low exposure limit. By contrast, fluorinated ethers show a linear increase in desorption yield with exposure. Figure 3 shows the desorption yield vs. exposure for diethyl ether and perfluorodiethyl ether on Ru(001). The desorption yield is determined by integrating the area under the thermal desorption spectrum. It is clear that the diethyl ether desorption yield increases slowly for exposures less than ca. 0.4 L. When the γ-state begins to fill, at ca. 0.4 L, the desorption yield increases at a faster rate. In contrast, the desorption yield of perfluorodiethyl ether varies linearly with exposure over the entire exposure range. Similar results are obtained for the other compounds listed in Table 1. The change in slope for hydrogenated ethers, illustrated in Fig. 3, suggests that a fraction of the chemisorbed molecules decompose rather than desorb. As a result, the number of hydrogenated molecules which <u>de</u>sorb from the surface in the low-exposure regime is less than the number which <u>ad</u>sorb, leading to a relatively low desorption yield.

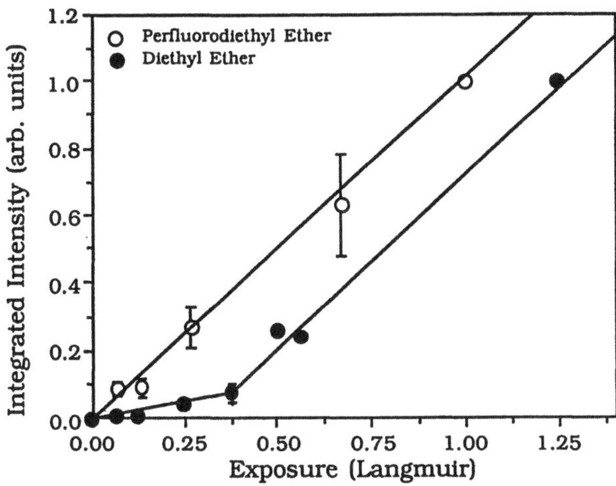

Figure 3. Desorption yield vs. exposure for diethyl and perfluorodiethyl ether on Ru(001).

The stability of chemisorbed fluorinated ethers, relative to hydrogenated ethers, is evident from the experimental facts discussed above. The decomposition mechanism of ethers is probably similar to the decomposition mechanism of alkoxides. The rate-determining step in the decomposition of surface ethoxide on Ni(111) is C-H bond breaking at the α-carbon, which occurs at ca. 260 K [30]. A similar decomposition mechanism is reported for methoxide adsorbed on Ru(001), where C-H bond cleavage occurs at 220 K [16].

If a decomposition mechanism similar to that for alkoxides operates for ethers, we expect decomposition of hydrogenated ethers to occur around 220 K on Ru(001). Chemisorbed diethoxymethane and diethoxyethane both remain on the surface up to this temperature. Decomposition via C-H bond breaking at an α-carbon is, therefore, a reasonable decomposition mechanism for these diethers.

There are good reasons, given below, to believe that the energy barrier for decomposition of the fluorocarbons exceeds that of the hydrocarbons. However, even if we assume for the moment that this barrier is the same for C-F and C-H bonds, the thermal desorption experiments could understandably induce a smaller extent of decomposition for the fluorocarbons than the hydrocarbons. This is because the fluorinated ethers desorb at lower temperatures, i.e. they simply may not stay in contact with the metal to temperatures high enough to initiate C-F bond breaking.

In addition to this effect, however, the barrier to breaking the C-F bond almost certainly exceeds that of the C-H bond. First, the C-F bond is simply stronger. For instance, the C-F bond in C_2H_6 (480 kJ/mole [31]) is stronger than the C-H bond in C_2F_6 (406 kJ/mole [32]). Second, there may be a higher energy barrier preventing the C-F bond from approaching the surface. Due to repulsion between electron-rich fluorine atoms and the metal surface, the alkyl side chains of a fluorinated ether may not approach the surface as closely as the alkyl groups of a hydrogenated ether, resulting in less decomposition.

4.2. Monoethers vs. Diethers

Since diethers have two functional groups, a variety of bonding
configurations are possible. One or both oxygen atoms can bond to the
surface and the alkyl groups can approach the surface closely or remain
far away. Diethers can, in principle, bond to the atomically-smooth
Ru(001) substrate via both oxygen atoms without introducing intramolecular
strain. Such $\eta^2(O,O)$-bonded molecules would form adsorption bonds with
strengths on the order of 80 kJ/mol as a first approximation.
Alternatively, only one oxygen atom could bond to the surface, perhaps for
entropic or electronic reasons. Such an $\eta^1(O)$-configuration should have
an ether-surface bond strength comparable to that of a monoether.

We find that the addition of a second ether linkage increases the
ether-surface bond strength (see Table 1), but diether-surface bonds are
less than twice as strong as monoether-surface bonds. This suggests an
$\eta^1(O)$-bonding configuration for the diethers on Ru(001). We attribute the
increase in bond strength of diethers over monoethers to the extra
methylene linkages in the diethers, which can also interact with the
surface.

5. Conclusions

We have studied the interaction between prototypical lubricant molecules
and metal surfaces with surface science techniques. We find that
<u>fluorinated</u> ethers bond more weakly to atomically smooth ruthenium surfaces
than analogous <u>hydrogenated</u> ethers. Diethers appear to bond through only
one ether linkage, since the bond strengths of diethers are not double
those of monoethers. Between 0.04 and 0.17 monolayers of hydrogenated
ethers (both monomers and dimers) decompose, while the fluorinated
compounds are very stable toward decomposition. These results suggest that
the surface properties of monomeric ethers, both hydrogenated and
fluorinated, serve as good indicators for the surface properties of
oligomers.

6. Acknowledgements

We sincerely thank Dr. Tom Bierschenk of Exfluor Research Corp. for
providing the perfluorodiethoxyethane. This research is supported by the
Director for Energy Research, Office of Basic Energy Sciences. Ames
Laboratory is operated for the U.S. Department of Energy by Iowa State
University under Contract No. W-7405-ENG-82.

7. References

1.   <u>Krytox Technical Bulletin L6</u>, Dupont Co.
2.   Y. Ohshka, T. Tohzuka, S. Takai: Eurpoean Patent Application No.
      84116003.9, Dakin Industries, Ltd. December 20, 1984.
3.   D. Sianesi, R. Fontanelli: U.S. Patent Number 3,665,041, Montefluos
      Inc. May 23, 1972.
4.   W. Morales, D. H. Buckley: Wear <u>123</u>, 345 (1988).
5.   W. R. Jones, K. J. L. Paciorek, J. H. Nakahara, M. E. Smythe, R. H.
      Kratzer: Ind. Eng. Chem. Res., <u>26</u>, 1930 (1987).
6.   W. R. Jones, Jr., K. J. L. Paciorek, D. H. Harris, M. E. Smythe, J.
      H. Nakahara, R. J. Kratzer: Ind. Eng. Chem. Prod. Res. Dev., <u>24</u>, 418
      (1985).
7.   W. R. Jones, K. J. L. Paciorek, T. I. Ito, R. H. Kratzer: Ind. Eng.
      Chem. Prod. Res. Dev., <u>22</u>, 166 (1983).

8.   M. M. Walczak, P. A. Thiel: J. Am. Chem. Soc., _109_, 5621 (1987).
9.   N. R. Avery, Langmuir, _1_, 162 (1985).
10. L. Ng, J. G. Chen, P. Basu, J. T. Yates, Jr.: Langmuir, _3_, 1161 (1987).
11. M.M. Walczak, P.A. Thiel: in preparation.
12. P.A. Redhead: Vacuum, _12_, 203 (1962).
13. A.B. Anton, N.R. Avery, B.H. Toby, W.H. Weinberg: J. Am. Chem. Soc., _108_, 684 (1986).
14. M.A. Henderson, P.L. Radloff, J.M. White, C.A. Mims: J. Chem. Phys., _92_, 4111 (1988).
15. M.A. Henderson, P.L. Radloff, C.M. Greenlief, J.M. White, C.A. Mims: J. Chem. Phys., _92_, 4120 (1988).
16. J. Hrbek, R.A. dePaola, F.M. Hoffmann: J. Chem. Phys., _81_, 2818 (1984).
17. N.R. Avery, B.H. Toby, A.B. Anton, W.H. Weinberg: Surf. Sci., _122_, L574 (1982).
18. A.B. Anton, J.E. Parmenter, W.H. Weinberg: J. Am. Chem. Soc., _108_, 1823 (1986).
19. E.D. Williams, W.H. Weinberg: Surf. Sci., _82_, 93 (1979).
20. P. Feulner, D. Menzel: Surf. Sci., _154_, 465 (1985).
21. M.Y. Chou, J.R. Chelikowsky: Phys. Rev. Lett., _59_, 1737 (1985).
22. H. Luth, G.W. Rubloff, W.D. Grobman: Surf. Sci., _63_, 325 (1977).
23. B.A. Sexton, A.E. Hughes: Surf. Sci., _140_, 227 (1984).
24. K.D. Rendulic, B.A. Sexton: J. Catalysis, _78_, 126 (1982).
25. P.A. Thiel, T.E. Madey: Surf. Sci. Reports, _7_, (1987).
26. T.E. Madey, J.T. Yates, Jr.: Surf. Sci., _76_, 397 (1978).
27. F.M. Hoffmann, T.H. Upton: J. Phys. Chem., _88_, 6209 (1984).
28. A.D. Mitchell, L.C. Cross, Eds.: _Tables of Interatomic Distances and Configurations in Molecules and Ions_, (Burlington House, London 1958).
29. R.D. Chambers: _Fluorine in Organic Chemistry_, (John Wiley and Sons, New York 1973).
30. S.M. Gates, J.N. Russell, Jr., J.T. Yates, Jr.: Surf. Sci. _171_, 111 (1986).
31. M. Stacey, J.C. Tatlow, A.G. Sharpe: _Advances in Fluorine Chemistry_, Vol. 2 (Butterworths, London 1961).
32. V.I. Vedeneyev, L.V. Gurv'ich, V.N. Kondrat'yev, V.A. Medvedev, Y.L. Frankevich: _Bond Energies, Ionization Potentials and Electron Affinities_, (Edward Arnold Ltd., London 1966).

MeV-Ion Induced Surface and Interface Chemistry *

P.F. Lyman[1] and L.E. Seiberling[2]

[1]University of Pennsylvania, Department of Physics,
 Philadelphia, PA 19104, USA
[2]Present address: University of Florida, Department of Physics,
 Gainesville, FL 32611, USA
*Supported by the NSF [DMR-8896262]

It is now well established that MeV-ion irradiation can greatly improve the adhesion of a thin film to a substrate. Several proposed mechanisms involve changes in an interfacial contaminant or oxide layer produced by electronic excitation during passage of the energetic ion through the interface. We have investigated the changes produced in a native oxide of Si(100) by bombardment with 5.9 MeV Be ions. The technique of transmission ion channeling was used to measure H, C, O and non-registered Si concentrations in the native oxide as a function of ion dose. The possibility of order in the native oxide was also investigated. We found that most of the H incorporated in the native oxide was randomly distributed on the surface, and rapidly desorbs due to electronic excitation during ion bombardment. The dose required to desorb roughly 30% of the H corresponds well to the dose required to significantly improve the adhesion of a metal layer over a native oxide on Si. A component of H was discovered that is ordered with respect to the silicon lattice. This H is not affected by ion bombardment and probably plays no role in adhesion. Evidence of bonding changes in O and Si during ion bombardment were also observed.

1. Introduction

It is now well established that MeV-ion irradiation can substantially improve the adhesion of a thin film to a substrate [1]. One of the proposed mechanisms is the formation of new chemical bonds across the interface [2,3]. The enhanced adhesion has been shown to be associated with the loss of energy by the ion to electronic scattering processes rather than to direct collisions with atoms at the interface [4,5]. It is thought that the disturbance of the electronic system could lead to broken bonds, and that the system could then relax into a different bonding configuration.

A physical manifestation of changes in bonding would be slight changes in the position and/or changes in the abundance of atoms at the interface. A system for which MeV-ion enhanced adhesion has been studied under a variety of conditions, and one that is important technologically is metal overlayers on a silicon substrate containing a thin native oxide. The silicon/native oxide system is particularly well suited to study using the technique of

Springer Series in Surface Sciences, Vol. 17
Adhesion and Friction   Editors: M. Grunze and H.J. Kreuzer
© Springer-Verlag Berlin, Heidelberg 1989

transmission ion channeling, because high-quality, thin single crystal samples can readily be produced. We have used transmission channeling to investigate movement, adsorption or desorption of atoms (H, C, O and non-registered Si) comprising the native oxide of silicon during ion bombardment. The hope was that these atomic displacements could be correlated to new bonds being formed within the native oxide.

Our results indicate that hydrogen exists in at least two bonding configurations prior to ion bombardment. One configuration leaves the hydrogen in an ordered array with respect to the silicon lattice, indicating that this hydrogen is bonded directly to the top silicon layer, or, perhaps to oxygen bonded directly to silicon. This hydrogen appears to be unaffected by ion bombardment. The other configuration is randomly positioned with respect to the silicon lattice, and rapidly desorbs during ion bombardment. This hydrogen may be incorporated in a loosely-bound hydrocarbon layer on top of the oxide. The data also indicate that Si and O are displaced initially during ion bombardment, but that no further movement occurs after a critical dose is achieved. No changes in C are observed with ion dose. We propose that in systems such as metal layers evaporated onto chemically cleaned silicon, a loosely adsorbed hydrocarbon layer inhibits adhesion. Ion bombardment releases hydrogen, and the bonds thereby freed reform across the interface, knitting the metal overlayer to the substrate.

2. Experimental Procedure

The experiments were performed on n-type device-grade commercially polished Si(100) wafers. Boron doping was used with a selective chemical etch to produce free-standing, single-crystal silicon films about 800 nm thick [6]. These samples were then cleaned according to the method of SHIRAKI [7]; this process is comprised of repeated chemical oxidations and HF etches. After the final etch in dilute HF, the samples were immediately placed in a load-lock chamber, and the chamber was evacuated. The samples could later be transferred _in vacuo_ from the holding chamber to a previously baked-out UHV scattering chamber.

A beam of 5.9 MeV Be^{++} ions was produced by the University of Pennsylvania FN tandem Van de Graaff accelerator and collimated to 0.1° angular divergence. This beam was directed along the <100> crystallographic direction normal to the sample surface. The unscattered portion of the beam passed through the sample and was collected in a Faraday cup and integrated to record the ion dose. Detectors were placed at scattering angles of 18°, 50°, and 85°, as depicted in Fig. 1a. The detector at 18° was covered by a foil which stopped all ions but energetic protons, and was used to measure the recoiling H atoms ejected from the sample. The grazing exit angle detector (85°) was used to measure ions scattered from the non-registered Si atoms at the beam-exit (back) surface; these events occur at higher energy than (and are thus distinguishable from) events originating from atoms

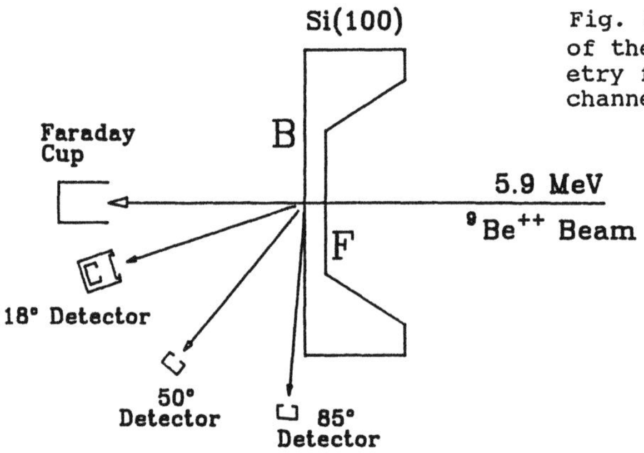

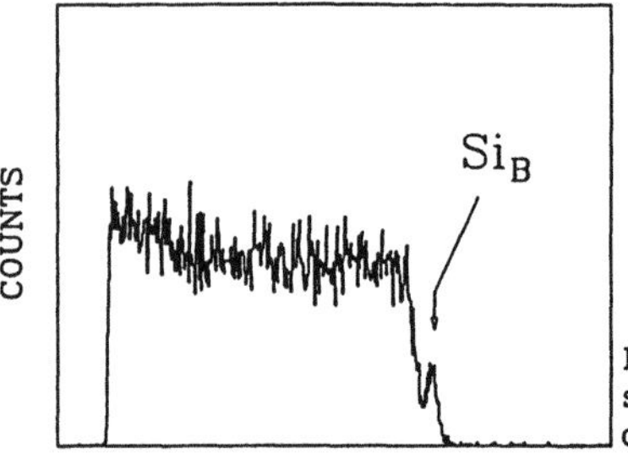

Fig. 1a Schematic diagram of the scattering geometry for the transmission channeling experiment

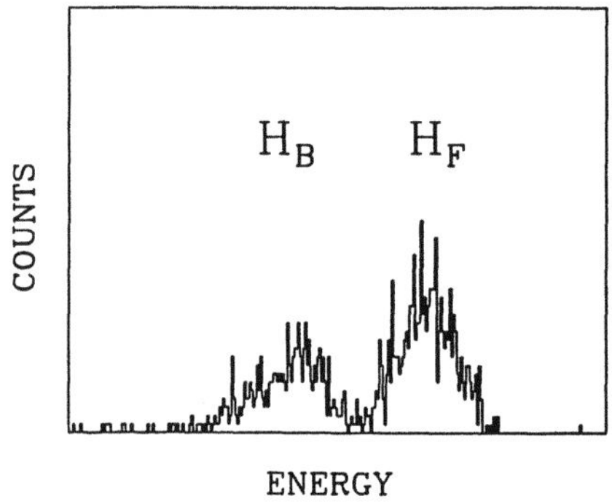

Fig. 1b Representative spectrum of hydrogen recoils from the 18° detector. The subscripts refer to the back (B) and front (F) surface of the sample

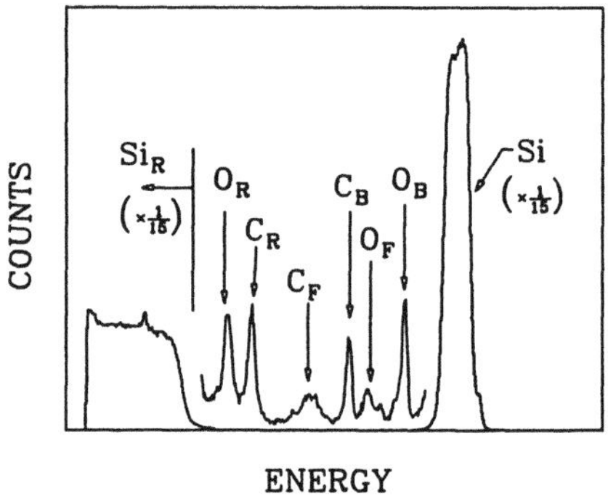

Fig. 1c Representative spectrum of Be ions scattered into the 50° detector. Peaks due to elastic recoils appear at low energies

Fig. 1d Representative spectrum of Be ions scattered into the 85° detector. The peak due to non-registered Si on the back surface

on bulk lattice sites due to the channeling effect. Finally, the detector at 50° was used to measure scattering and recoil events from intermediate mass elements, C and O in this case. Representative spectra from these detectors are shown in Figs. 1b, c and d, respectively.

The experiments were conducted in the channeling direction to reduce background from the bulk silicon, and to allow the detection of non-registered silicon as noted above. After locating the channeling direction, the sample was translated to a fresh beam spot and consecutive doses of approximately 5.6×10^{15} ions/cm^2 were administered. Spectra from each dose were analyzed to yield concentrations of H, C, O and non-registered Si. The peak areas from the scattering spectra were converted to surface concentrations using the scattering cross sections and the detector solid angles. The Rutherford cross section was used for Si, C, and O events, while the cross section for the H recoils was measured [8] to be 1.82 times larger than the Rutherford value for Be ions of this energy. The areal densities were calculated assuming that the species were distributed randomly across the surface.

3. Results

For comparison to the present study of the Si(100) surface, we show results of earlier work on the Si(110) surface. Figure 2 depicts the surface concentration of Si in the (110) native oxide that has moved laterally from its bulk position, as a function of ion dose [9]. The linear rise followed by a saturation after a dose of about 5×10^{16} ions/cm^2 has been reproduced over a wide range of pressures and beam current densities. Figures 3a and 3b, respectively, depict the Si(110) [2] and Si(100) concentrations of Si and O on the back (B) surface, as a function of ion dose. The circles represent O and the crosses represent non-registered Si. The solid line on Fig. 3a is a fit to the data of Fig. 2. The data in Fig. 3b were fit with a curve comprised of a linear rise followed by a constant line, in accordance with typical behavior observed for the (110) samples. The additional scatter in Fig. 3b is due to poorer counting statistics. The

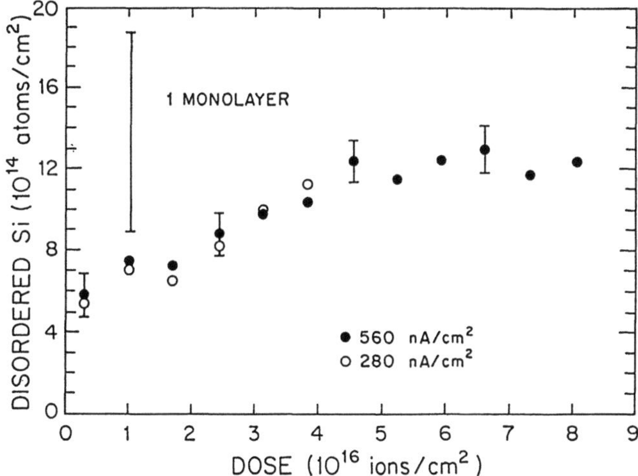

Fig. 2 The surface concentration of non-registered Si as a function of ion dose for a (110) sample. The ambient pressure was 2×10^{-7} Torr. Two current densities were run with identical results

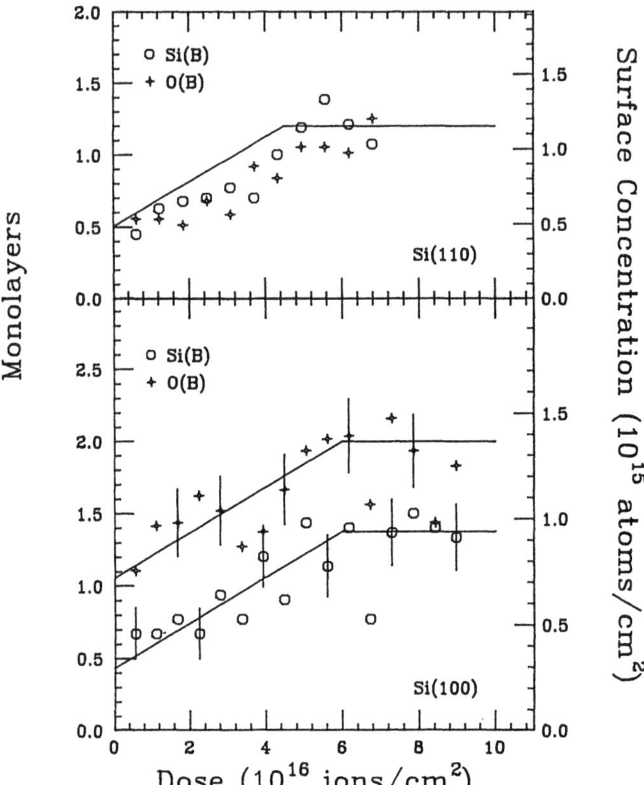

Fig. 3a The concentration of Si and O on the back surface (B) as a function of ion dose, for a Si(110) sample. The solid line os a fit to the data of Fig. 2

Fig. 3b The concentration of Si and O on the back surface (B) as a function of ion dose, for a Si(100) sample. The curves are a fit assuming a linear rise followed by a constant value of the concentration

curves provide a reasonably good fit to the data, and indicate that the O signal rises at the same rate as the Si signal.

Figures 4a and 4b, respectively, depict surface concentrations on Si(110) [2] and Si(100) of H and C, as a function of dose. The C coverages remain roughly constant at 2.9 and 2.4 × 10^{15} atoms/cm^2. In the (110) experiment, the H concentration on the back surface, H(B), could not be clearly resolved from that on the front, H(F), so the average is shown by the circles. The solid curve is a best fit of these data to an exponential decay with a constant background, $Ae^{-\sigma D} + B$. The desorption cross section was 6.5 × 10^{-17} cm^2. In the (100) experiment, the H(B) data are plotted with crosses and the H(F) data are plotted with circles. The solid curves employ the same desorption cross section as the (110) data, but have backgrounds adjusted to best match the data.

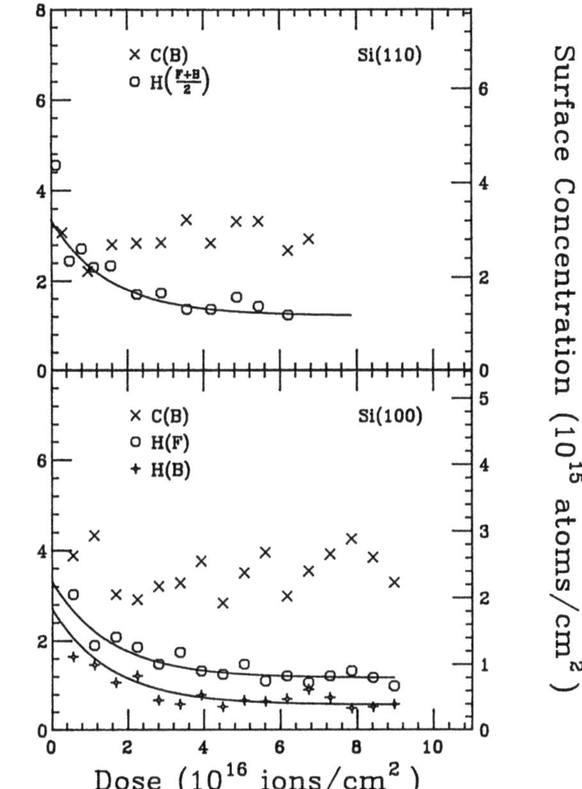

Fig. 4a The concentration of C on the back surface and the average of the H concentration on the back and front surfaces as a function of ion dose for a Si(110) sample. An average was used for hydrogen because the front and back peaks could not be resolved. The solid curve is a best fit of these data to an exponential decay with a constant background.

Fig. 4b The concentration of C on the back surface and H on the front and back surfaces as a function of ion dose for a Si(100) sample. The solid curves are similar to the curve in Fig. 4a, with the same desorption cross section, but with the background adjusted to best match the data.

The striking feature of the H data is the constant difference of 1.2 monolayers (ML) between the coverage measured on the front face and that on the back, over the entire range of ion fluence. This can be due either to a larger concentration of H on the front surface, or it can be due to channeling effects. If, for example, the H on the back surface were located along the rows of silicon atoms, then a decreased yield would be expected for H(B), because these atoms would be hidden from the ion beam. In order to distinguish these cases, the ion beam was directed in a random incident direction so that the effects of channeling were eliminated. The coverages measured on the front and back surfaces were found to be the same within experimental error. We then conclude that the H left on the surface at the saturation dose was ordered, with a channeling yield of approximately 0.5.

4. Discussion

Some insight into the role of electronic stopping power in MeV ion-enhanced adhesion may be gained from examining the H data. The initial coverage of H(F) was found to be 3.3 ML, but fell to 1.2 ML at saturation. The difference between the H(F) and the H(B) curves remains fairly constant throughout the ion bombardment. This implies that the H that desorbed was equally discernable to both the channeled ion flux at the back surface and the uniform ion flux at the front surface; this means that the H that desorbed was distributed randomly across the surface. The cross section for H desorption noted above corresponds to an ion impact parameter of nearly a Bohr radius. This large cross section is a clear hallmark of an electronic process. The rapid depletion of H in the oxide layer would free bonds that could subsequently reform with other atoms present at the interface. The dose required for significant improvement of adhesion for most metals on silicon is less than 1×10^{16} ions/cm^2 [10]; this corresponds to roughly a 30% loss of H from our sample. It is interesting to note that the H atoms that do not desorb exhibit order with respect to the silicon lattice. The fact that they persist through the adhesion threshold seems to imply that they do not play a role in this phenomenon.

In the (100) data, the O coverage increases from 1 ML to 2 ML and then saturates. An interesting problem is the source of the additional O. The earlier work was done at pressures around 3×10^{-9} Torr, and it was assumed that O was taken up from the ambient vacuum. The present study was conducted with pressures one order of magnitude lower, and it seems unlikely that there would be a sufficient supply of O to the surface to account for the increase. Adsorption from the vacuum would require one half of our background pressure of 4×10^{-10} Torr to be produced by O-containing molecules, even assuming a sticking coefficient of unity. An alternative explanation is that 2 ML of O was present on the surface at the commencement of the ion bombardment, but it had a channeling yield of about one half. The ion bombardment then induced chemical changes which distributed the O randomly over the surface. The clear signs of order in the H data make it plausible that the O is initially ordered. Furthermore, order has recently been reported in the native oxide of silicon [11], lending credence to this scenario.

The O and Si coverages increased at the same rate and saturated at the same dose in Si(100), indicating that they are reacting with each other under the influence of ion bombardment. We would like to argue that the increase in non-registered Si is a result of chemistry induced by electronic excitations, rather than damage resulting from knock-on collisional processes. Figure 5 compares the

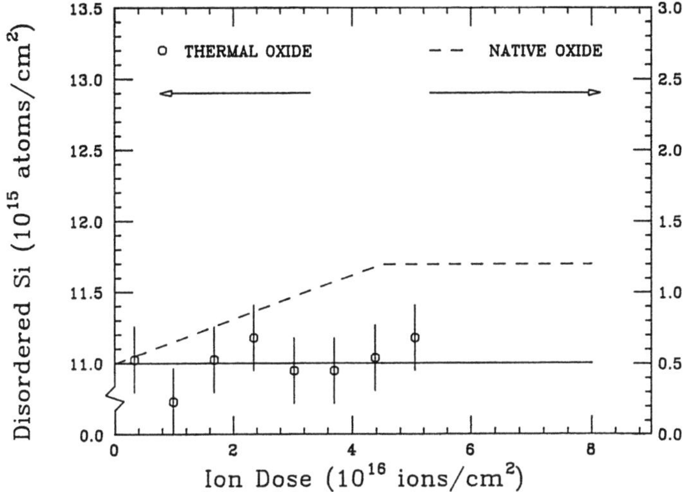

Fig. 5 The surface concentration of Si as a function of ion dose for both a thin, thermally grown oxide and a native oxide. The left and right axes have different offset values, but the same scale.

behavior of a thin, thermally-grown oxide [d] to a native oxide under ion bombardment. The left and right axes have different offset values, but the same scale. The lack of a rapid initial increase in non-registered Si for the thermal oxide argues against a knock-on process being responsible for the production of the non-registered Si. Knock-on production of Si disorder would become more likely with increasing disorder. For the same reason, it is hard to see how knock-on process could lead to a saturation in the native oxide disorder. However, chemistry induced by electronic processes could easily be imagined to lead to saturation.

5. Conclusions

We have investigated the changes produced in the native oxide of silicon during bombardment with MeV ions. We find that most of the H initally incorporated in the native oxide is randomly distributed, and rapidly desorbs due to electronic excitation during ion bombardment. The dose required to desorb roughly half of the H corresponds well to the dose required to significantly improve the adhesion of a metal layer over a native oxide on Si. We propose that H released during ion bombardment frees up bonds that can reform across the interface, thereby improving the adhesion. A component of H was discovered that is ordered with respect to the silicon lattice. This H is not affected by ion bombardment and probably plays no role in adhesion. Evidence of bonding changes in O and Si during ion bombardment was also seen.

References

1 J. E. Griffith, Y. Qiu and T. A. Tombrello, Nucl. Instrum. Meth. **198**, 607 (1982).

2 L. E. Seiberling, Nucl. Instrum. Meth. **B24/25**, 526 (1987).

3 C. J. Sofield, C. J. Woods, C. Wild, J. C Riviere and L. S. Welsh, Mat. Res. Soc. Sypm. Proc. **25**, 197 (1984).

4 T. A. Tombrello, Nucl. Instrum. Meth. 230, 555 (1984).

5 I. V. Mitchell, J. S. Williams, P. Smith and R. G. Elliman, Appl. Phys. Lett. **44**, 193 (1984).

6 N. W. Cheung, Rev. Sci. Instrum. **51**, 1212 (1980).

7 A. Ishizaka, K. Nakagawa and Y. Shiraki, 2nd Int. Symp. on Molecular Beam Epitaxy, Tokyo, 183 (1982).

8 F. S. Mozer, Phys. Rev. **B 104**, 1386 (1956).

9 L. E. Seiberling and R. L. Headrick, **Surface and Colloid Science in Computer Technology**, K. L. Mittal, ed., Plenum Press, NY, 235 (1987).

10 M.H. Mendenhall, Ph.D. thesis, Calif. Inst. of Tech., 1983.

11 A. Ourmazd, D. W. Taylor, J. A. Rentschler and J. Bevk, Phys. Rev. Lett. **59**, 213 (1987).

MeV-Ion Induced Changes in Thin Film Adhesion *

*T.A. Tombrello***

Schlumberger Doll Research, Ridgefield, CT 06877, USA
**On leave from the California Institute of Technology
*Supported in part by the NSF [DMR86-15641]

It has been known for some time that irradiation by MeV ions can greatly improve the adhesion of a thin film to a substrate. A number of different mechanisms have been shown to be involved: surface cracking of the substrate; the relaxing of substrate surface strain; the breakup of contaminant and oxide layers between substrate and film; and the formation of new chemical bonds between film and substrate atoms. In this paper I shall concentrate on the role played by the secondary electrons that are generated in the ion's passage. A sufficient number of such electrons are produced by each ion that it is possible to break simultaneously a number of chemical bonds in close spatial proximity, thus providing an underlying route to the mechanisms listed above.

1. Introduction

MeV-ion irradiation of many thin film-substrate systems has been shown to improve dramatically the degree to adhesion. [1,2] These observations show that the adhesion enhancement is correlated with the energy lost by the ions to the electronic degrees of freedom in the materials - not to direct collisions with the target atoms. This conclusion was put on an even more solid footing by experiments that demonstrated that similar results could be obtained by keV electron bombardment, where no direct atomic displacement was possible. [3] Since atomic mixing could not be a major component of the process, it was suspected that breaking of chemical bonds by the incident ion allowed new bonds to form across the interface. The formation of these bonds might cause the movement of interfacial atoms, but only a few atomic layers were likely to take part. This was confirmed in the transmission channeling experiments of HEADRICK and SEIBERLING. [4]

As more data were accumulated it became clear that a variety of mechanisms were involved in the modification of adhesion by MeV-ion and keV-electron bombardment. These mechanisms include: surface cracking of the substrate - possibly involving electrostatic effects at the crack edges [5]; breaking up of oxide or contaminant layers between film and substrate [6,7]; changes in the mechanical stresses at the interface [6,8,9,10]; and changes in chemical bonding [11].

Most of this work concentrated either on systems where the intrinsic adhesion was low (e.g., Au films on teflon or silica) or where the establishment of a strong mechanical and electrical contact was important (e.g., conductors on semiconductors). The dependence of the effect on ion type and bombardment fluence provided the first indication that the mechanism involved the excitation of electrons in the material; however, systematic measurements for wide variations of even these variables were not made.

Springer Series in Surface Sciences, Vol. 17
Adhesion and Friction   Editors: M. Grunze and H.J. Kreuzer
© Springer-Verlag Berlin, Heidelberg 1989

In the following sections I suggest a major mechanism for this adhesion enhancement effect that allows model calculations; however, the lack of extensive and systematic measurements makes this a speculative proposal that is tied tenuously to the existing data. Thus, I advance these ideas mainly as a stimulus for further work in this area and in the hope that they may provide a framework to focus your attention for the initial phase of these investigations.

2. Proposed Mechanism

Although direct collisions with individual atoms probably play a minor role in surface stress relaxation [9,10], the major contributor to the adhesion enhancement comes from the interaction of the ion with electrons [2,3]. This was first demonstrated by the dependence of adhesion test thresholds on the energy loss of the ion in the material (dE/dx). [2] It is well known that a large portion of this energy goes into the production of a shower of secondary electrons that originate on the ion's path and extend tens of Ångstroms away from it. For example, at the exit surface of a thin carbon film that is irradiated with 10 MeV ^{35}Cl ions there is a yield of 10^2 secondary electrons. [12] The fact that there are so many energetic secondary electrons present simultaneously allows processes to occur that would be extremely improbably if they had to be accomplished by sequential events. One such example is provided by the destruction of living cells by ionizing radiation, where a large biological molecule is likely to be killed only if it is hit simultaneously by a number of these secondary electrons. [13] Thus, the secondary electrons allow both the irradiation of a large area and the simultaneous breaking of a large number of bonds within it.

Since the number of secondary electrons produced by MeV ions is proportional to dE/dx, how can one tell whether the secondary electrons or merely the total deposited energy is responsible for a given effect? As the energy of the ion increases, the energies of the secondary electrons also increase and they have longer ranges in the material. Thus, although the total energy of the secondary electrons is proportional to dE/dx, the energy deposited by those electrons in a given area perpendicular to the ion's path will peak earlier than dE/dx and decrease faster. This behavior under MeV ion irradiation has been observed and explained in terms of such a model for the erosion of dielectric materials [14], for the desorption of large organic molecules [15], and for the changes of resistivity induced in amorphous carbon films [16].

The model developed by Hedin et al. to explain the desorption of large organic molecules from surfaces by MeV ions [15] can be adapted easily to the adhesion enhancement effect discussed herein. The secondary electrons generated in the film shower upon the interface, allowing an extended area to have many bonds broken simultaneously. This bond breaking can produce many of the effects that were described in the introduction: breaking up of oxide or contaminant layers; breaking substrate or film bonds so that new bonds can form across the interface; even modifying the state of mechanical stress at the interface. Once one specifies the area and the number of bonds that have to be broken simultaneously then even the absolute value of the cross section (proportional to the reciprocal of the threshold fluence) is predicted.

3. Comparison of the Model with Data

As we stated in the introduction, the ion-induced-adhesion enhancement has never been studied systematically for the variables that are important to confirm the model described in the previous section. To do that we would need to look at a single film-substrate system for a wide range of incident ion velocities (v) for several ions with appreciably different atomic numbers (Z). This would allow the v and Z dependence of the model to be checked. A detailed comparison with the model predictions of absolute adhesion thresholds with ion fluence is much harder to achieve. Although one can make reproducible threshold fluence measurements for a given film-substrate preparation technique, any change in preparation, cleaning, or deposition procedures shifts the absolute thresholds obtained. [2] This is not unexpected, especially when oxide layers or contaminants at the interface play a role. We shall give absolute values; however, we do not feel that they have any special significance at this point - although they may be useful in the future when better data are available.

Our first example is taken from the experiments of Wie et al., in which 350Å Au films on vitreous silica (amorphous SiO_2) were irradiated with ^{35}Cl ions at MeV energies. [3] Figure 1 shows the reciprocal of the ion fluence at which the adhesion just passes the Scotch tape test versus the ion velocity (in units of c). Also shown is dE/dx, which peaks at a higher velocity and varies more slowly. We have also given the cross section calculated from the model [15] for a region 10Å across where 8 bonds are broken simultaneously. The peak in the calculated curve does not depend significantly on the number of bonds broken; the fact that the data are so sharply peaked seems to imply that we need to assume more bonds are broken; 11-12 would give about the right width. However, this would drop the cross section, which is already a factor of twenty lower, by another factor of five.

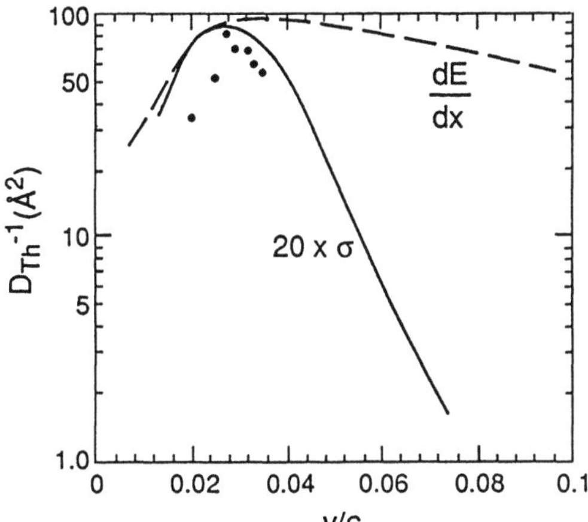

Figure 1: The reciprocal of the ^{35}Cl ion fluence (in $Å^2$) required for a 350Å Au film on amorphous SiO_2 to pass the Scotch tape test versus the ratio of the ion velocity to the velocity of light. [5] Shown as a dashed curve is dE/dx for the ions (arbitrary units); at the peak dE/dx is approximately 10 $MeVcm^2/mg$. The solid curve is calculated using the model of Hedin et al., [15] for breaking eight bonds simultaneously in a target region of 10Å size; the calculated value has been multiplied by a factor of 20.

In [2] it was established that for 500Å Au films on Ta (probably with a thin oxide layer) the fluence necessary to pass the Scotch tape test varied as (dE/dx)[-1.6]. These data were obtained with a variety of ion types, so we are unable to make the same comparison as in our first example. In Fig. 2 we show these data and the model [15] predictions for 4 and 8 bonds broken simultaneously. In the latter case the agreement is not too bad for dE/dx$\geq$ 1 MeV cm^2/mg. Again the model cross sections are lower - in this case by a factor of three.

Having exhausted the data that can be compared directly, our last example does not allow a straightforward model prediction. Hull et al. irradiated 900 Å Au films on glass substrates as a function of keV electron energy and measured the adhesion strength by pull tests. [17] Their data are shown in Fig. 3, where the incident electron energies are given - not their value at the interface. The concept of electron range is not well defined, but 900 Å of Au is roughly equivalent to the projected "range" of a 7 keV electron. We show the energy loss at the interface provided in [17] for reference; the dotted curve gives our crude modification of this curve to show how the density of deposited energy from the secondary electrons will vary. As we indicated earlier, the key factor to notice is how the adhesion drops faster than the total deposited energy.

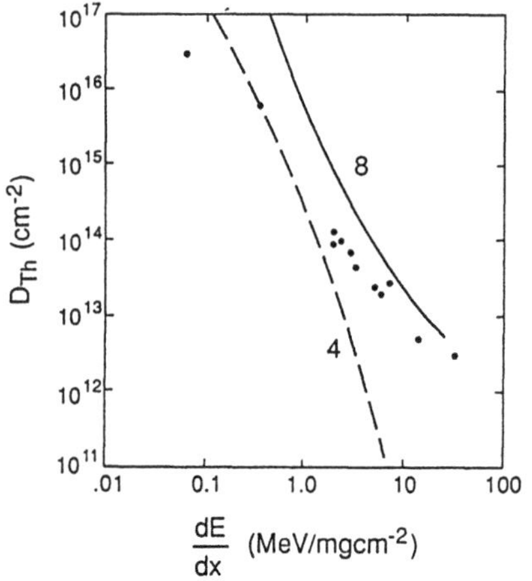

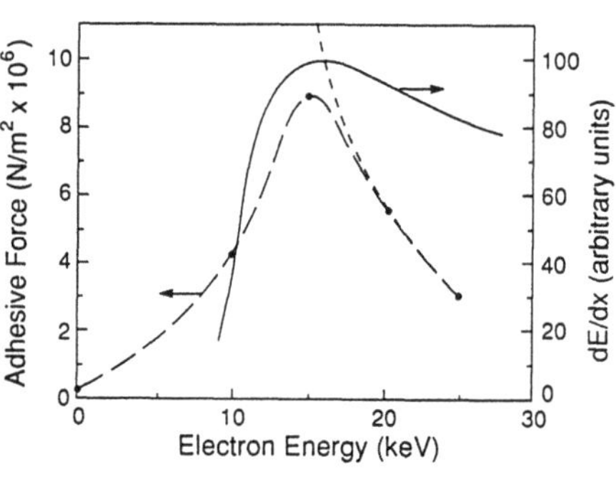

Figure 2: The threshold fluence to pass the Scotch tape test for 500Å Au films on Ta is plotted versus the energy loss of the bombarding ion in the film. The solid curve corresponds to the model of Hedin et al. [15] for breaking 8 bonds in the target region (10Å) simultaneously; the dashed curve is for 4 bonds broken.

Figure 3: The measured adhesive force as a function of electron energy at the surface of a 900Å Au film on glass for a fluence of 10^{16} cm^{-2}. [17] The solid curve is dE/dx for the electrons. [17] The dotted curve is an estimate of the way dE/dx is modified by the increased cross sectional area over which the secondary electrons deposit their energy.

<u>4. Conclusion</u>

Previously, the single model that has been put forward to explain the enhancement of thin film adhesion by MeV-ion irradiation assumes only a dependence on dE/dx. [18,19] This model was not mechanistic, but used a calorimetric approach through the Richardson-Dushman equation to connect the deposited energy to a flow of electrons across the interface. This electron flow was in turn equated to the broken bonds that modified the adhesion. In [18] these electrons were the secondaries from the ion's passage; in [19] they were from an equilibrated electron plasma. The two approaches gave similar results but with a slightly different dependence on dE/dx.

In the work reported here, I have set out to provide a mechanistic foundation for the effect that is also based on the production of secondary electrons by the ion and their subsequent breaking of chemical bonds in the region around the ion's path. To reduce this mechanism to its essence, the model of Katz [13] has been used for calculations with the formulae derived by Hedin et al. for molecular desorption [15]. This particular embodiment of the mechanism should not be regarded as its ultimate refinement - merely all that is justified by the amount of data that are available now for comparison.

The data of Wie et al. show enough of the features of the proposed mechanism (its peak is sharper and at a lower energy than dE/dx) to justify a few new experiments that are specifically designed to test these ideas. The two other examples give little detailed support to the model, but both confirm the general trends expected.

What is obviously needed is a carefully chosen film-substrate system where competing effects (e.g., interface stress modification) can be minimized. A likely candidate would be an Au film on a clean substrate having a thin native oxide, where there is low intrinsic adhesion. The mechanism would then be expected to involve the breaking of oxygen-substrate bonds, allowing the formation of stronger bonds between Au and substrate atoms. High-purity silicon represents an excellent choice for the substrate, but other choices like Al would also be suitable.

I hope that by making this proposal when few data are available, you will be provoked into setting out to prove me wrong. The only way that we are going to make something substantial out of this effect, rather than continuing to treat it as a curious novelty, will be to do systematic experiments. The framework provided by the mechanism proposed is intended as a start in this direction.

<u>References</u>

1. J.E. Griffith, Y. Qin, and T.A. Tombrello, Nucl. Instrum. Meth. <u>198</u>, 607 (1982)

2. T.A. Tombrello, Nucl., Instrum. Meth. <u>230</u>, 555 (1984) and Mat. Res. Soc. Symp. Proc. <u>25</u>, 173 (1984); M.H. Mendenhall, Ph.D. thesis, Calif. Inst. of Tech., 1983

3. I.V. Mitchell, J.S. Williams, P. Smith, and R.G. Elliman, Appl. Phys. Lett. <u>44</u>, 193 (1984)

4. R.L. Headrick and L.E. Seiberling, Appl. Phys. Lett. $\underline{45}$, 388 (1984)

5. C.R. Wie, C.R. Shi, M.H. Mendenhall, R.P. Livi, T. Vreeland Jr., and T.A. Tombrello, Nucl. Instrum. Meth. $\underline{B9}$, 20 (1985)

6. M. Carbucicchio, A. Valenti, G. Battaglin, P. Mazzoldi, and R. Dal Machio, Hyperfine Interactions $\underline{29}$, 1213 (1986)

7. H. Dallaporta and A. Cros, Appl. Phys. Lett. $\underline{48}$, 1357 (1986)

8. P.A. Ingemarsson, T. Ericsson, A. Gustavsson-Seidel, G. Possnert, B.U.R. Sundqvist, and R. Wäppling, in <u>Proc. of the 12th Int'l. Symp. of Hosei Univ. Appl. of Ion Beams in Materials Science</u> (Tokyo, Sept. 2-4, 1987) in press.

9. R.P. Livi, S. Paine, C.R. Wie, M.H. Mendenhall, J.Y. Tang, T. Vreeland Jr., and T.A. Tombrello, Mat. Res. Soc. Symp. Proc. $\underline{37}$, 467 (1985).

10. T.A. Tombrello, Mat. Res. Soc. Symp. Proc. (April 1988 meeting of the Materials Research Society) in press.

11. C.J. Sofield, C.J. Woods, C. Wild, J.C. Riviere, and L.S. Welsh, Mat. Res. Soc. Symp. Proc. $\underline{25}$, 197 (1984).

12. C.R. Shi, H.S. Toh, D. Lo, R.P. Livi, M.H. Mendenhall, D.Z. Zhang, and T.A. Tombrello, Nucl., Instrum, Meth. $\underline{B9}$, 263 (1985).

13. E.J. Kobetich and R. Katz, Phys. Rev. $\underline{170}$, 391 (1968) and R. Katz, Nucl. Track Detection $\underline{2}$, 1 (1978).

14. C.C. Watson and T.A. Tombrello, Rad. Eff. $\underline{89}$, (1985) 263.

15. A. Hedin, P. Håkansson, B. Sundqvist, and R.E. Johnson, Phys. Rev. $\underline{B31}$ 1780 (1985).

16. T.A. Tombrello, Nucl. Instr. Meth. $\underline{B/24/25}$, 517 (1987).

17. T.R. Hull, J.S. Colligon, and A.E. Hill, Rad. Eff. Express $\underline{1}$, 55 (1987).

18. T.A. Tombrello, Int. J. Mass Spec. Ion Phys. $\underline{53}$, 307 (1983).

19. R.S. Strokstad, P.M. Jacobs, I. Tserruya, L. Sapir, and G. Mamane, Nucl. Instrum. Meth. $\underline{B16}$, 465 (1986).

Dynamics of Adsorbates on Metal Surfaces

C. Mavroyannis

Laser Chemistry Group, Division of Chemistry, National Research
Council of Canada, Ottawa, Ontario, Canada K1A 0R6

A microscopic theory is presented concerning the excitation spectra of
neutral rare-gas atoms physisorbed on metal surfaces. Emphasis has been
given to the dynamic effects of the surface plasmons on the lifetimes of the
adsorbed atoms; the effect of the surface plasmons is to split the optical
spectra of the adsorbed atoms into the atomic-like and plasmon-like excitations. At low coverage and when the nonradiative process due to the damping
of the surface plasmons dominates, the relative intensities per atom for the
peaks of the symmetric and antisymmetric modes take positive and negative
values describing the physical processes of absorption and stimulated emission, respectively. The red (blue) shifted peak of the symmetric mode of the
higher (lower) excited state and the blue (red) shifted peak of the antisymmetric mode of the lower (higher) excited state of the atom cancel each other
out provided that the frequency profiles of their lineshapes nearly coincide.
This may be a possible explanation of the persistence-extinction phenomenon
that has been observed for a number of rare-gas metal substrate systems in
the low coverage limit, where it has been proposed that a charge-transfer
instability exists. Results of numerical calculations have been graphically
presented and discussed; the vanishing computed configurations of Xe on Ti
and on Au and the persistence of the configurations of Ar on Au and on Mg are
compatible with the experimental observations.

1. Introduction

Flynn and coworkers [1-3] have extensively examined the optical excitation
spectra of rare-gas atoms physisorbed on various metal surfaces as a function
of coverage. The data were obtained by the use of differential reflectance
methods with synchrotron radiation. The primary spectroscopic effort has
focused on the way the substrate modifies levels of the free adsorbate, and
on any newly formed excited configurations which arise through the interaction of the adsorbate atom with the metal surface. The striking changes in
the optical excitation spectra have been interpreted as due to the formation
of a neutral excited adsorbate configuration on some surfaces and an ionized
one on others [1-3]. In the low coverage limit, the peak in the spectrum
remains strong for some adatom-substrate combinations, while it completely
vanishes for others. For instance, the adsorption peaks of Xe on Al, Ti and
Au, and Kr on Au and Mg are virtually eliminated while those of Xe and Ar on
Mg, Kr and Ar on Al and Ar on Au persist [1-3]. Objections to the foregoing
interpretations of the optical data [1-3] have been proposed [4,5]. No
theoretical explanation has been formed [4,6] for the occasional disappearance of the excitonic component of the optical-absorption system at low
coverage. Recent studies [7-9] have taken into consideration the coupling of
the adsorbate to the surface plasmon modes thus, allowing for screening
interactions to participate.

The problem is formulated in Section 2, where the model Hamiltonian
describing the system in question is used to derive the expressions for the

appropriate Green functions of the system. The excitation spectra of the
symmetric and antisymmetric modes of the system have been considered in
Section 3 while nonradiative processes are considered in Section 4. The
computed spectra of excited Xe and Ar on various metal surfaces are graphically presented and discussed in Section 5. Concluding remarks are given in
Section 6.

2. The Model Hamiltonian

We consider a simple model [10-14], which consists of a plane boundary separating an empty space and a homogeneous isotropic metallic medium having a
dielectric function $\varepsilon(\omega)=1-\omega_p^2/(\omega^2+i\omega\Upsilon_{sp})$, where ω_p is the plasma frequency of
the metallic electrons and Υ_{sp} ($\Upsilon_{sp}<<\omega$) is the intrinsic damping of surface
plasmons [15]. A neutral atom A is fixed at $(0,0,R)$ with the z-axis taken
perpendicular to the plane $z=0$. The image of the atom A, which is designated
by B, will be situated at $(0,0,-R)$ from the plane in the medium with dielectric function $\varepsilon(\omega)$. Each electron $(x_i,y_i,R-z_i)$ of the atom A with charge e
will have an image $(x_i,y_i,-R+z_i)$ having a fictitious charge $-e\varepsilon_1$, where
$\varepsilon_1=\varepsilon_1(\omega)=[\varepsilon(\omega)-1]/[\varepsilon(\omega)+1]$. The atom A is assumed to have nondegenerate
electronic states, the ground and excited states denoted by $|0\rangle$ and $|\nu\rangle$ with
frequencies E_0 and E_ν, respectively. The electronic transitions $|0\rangle\leftrightarrow|\nu\rangle$ are
electric dipole allowed and the transition frequencies are denoted by
$E_{\nu 0}=E_\nu-E_0$; units in which $\hbar=1$ are used throughout. The Hamiltonian
describing the system consisting of the atom A interacting with its image B
at a distance 2R apart may be taken to be [10,11,13].

$$H = \sum_\nu E_{\nu 0}\left(b_{A\nu}^+ b_{A\nu}+b_{B\nu}^+ b_{B\nu}\right) + \sum_\nu V_{AB}^\nu \bar{b}_{A\nu}\bar{b}_{B\nu} + \sum_{\vec{k},\lambda} ck\, \beta_{\vec{k}\lambda}^+ \beta_{\vec{k}\lambda} + \frac{\bar{\omega}_p^2}{4} \sum_{\vec{k},\lambda} \frac{1}{ck}\, \bar{\beta}_{\vec{k}\lambda}^+ \bar{\beta}_{\vec{k}\lambda}$$

$$+ \frac{i\bar{\omega}_p}{2} \sum_{\vec{k},\lambda,\nu} \left[f_{0\nu}(\vec{k},\lambda)\frac{E_{\nu 0}}{ck}\right]^{1/2}\left(e^{-i\vec{k}\cdot\vec{R}}\bar{b}_{A\nu}+\varepsilon_1 e^{i\vec{k}\cdot\vec{R}}\bar{b}_{B\nu}\right)\bar{\beta}_{\vec{k}\lambda} , \qquad (1)$$

where $f_{0\nu}(\vec{k},\lambda)$ is the oscillator strength of the electronic transition
$|0\rangle\leftrightarrow|\nu\rangle$ and ω_p is the plasma frequency of the atomic electrons defined by
$\bar{\omega}_p^2=4\pi e^2 S/mV$, where S is the number of all optically active electrons of the
neutral atom A; and m and V are the electron mass and the volume of the
sample container, respectively. The operators $\beta_{\vec{k}\lambda}^+$ and $\beta_{\vec{k}\lambda}$, $\bar{\beta}_{\vec{k}\lambda}=\beta_{\vec{k}\lambda}+\beta_{\vec{k}\lambda}^+$, are
the boson creation and annihilation operators describing the electromagnetic
field with wavevector $\vec{k}$, frequency ck and transverse polarization $\lambda=1,2$. The
atomic excitation operators $b_{A\nu}^+, b_{A\nu}$ and $b_{B\nu}^+, b_{B\nu}$ are defined as $b_{A\nu}^+=\alpha_{A\nu}^+\alpha_{A0}$,
$b_{A\nu}=\alpha_{A0}^+\alpha_{A\nu}$, $b_{B\nu}^+=\alpha_{B\nu}^+\alpha_{B0}$, $b_{B\nu}=\alpha_{B0}^+\alpha_{B\nu}$, $\bar{b}_{A\nu}=b_{A\nu}+b_{A\nu}^+$, $\tilde{b}_{A\nu}=b_{A\nu}-b_{A\nu}^+$, where $\alpha_{Ai}^+,\alpha_{Bi}^+$
and α_{Ai},α_{Bi} are the Fermi creation and annihilation operators describing the

electron states $i=|0\rangle$ and $i=|\nu\rangle$ of the atom A and B, respectively; and the corresponding electron number operators are defined as $n_{Ai}=\alpha^{+}_{Ai}\alpha_{Ai}$ and $n_{Bi}=\alpha^{+}_{Bi}\alpha_{Bi}$.

The first term on the right-hand side of (1) represents the free atomic fields while the second term describes the Coulomb interaction between the atom A and its image B; the coupling function $V^{\nu}_{AB}=\bar{V}\varepsilon_1$, where $\bar{V}=\sum U^{AB}_{0\nu}(\vec{R})$ is the Coulomb interaction, which in the dipole-dipole approximation, is the Lennard-Jones energy of interaction [11,12]. The third and fourth terms represent the free electromagnetic field and the contribution arising from the square of the vector potential term of the electromagnetic field, respectively. The last term describes the interaction of the electromagnetic field with the electron states of the atoms A and B. All the interactions involved in the Hamiltonian (1) are weak perturbations and, therefore, any time-dependent perturbation method when appropriately chosen is suitable to study the required excitation spectra arising from radiative and nonradiative decay mechanisms. We shall utilize the retarded double-time Green functions whose properties as well as whose use in statistical physics can be found elsewhere [10].

The atom A and its image B interact through their dipole-dipole inter-actions and radiate to each other as well and, hence, due to the cooperative radiative interaction between the atom A and its image B two new states, the symmetric and antisymmetric (with respect to interchange between the atoms), are generated [16]. We are interested in calculating the excitation spectra due to the process $|0\rangle\leftrightarrow|\nu\rangle$ for the symmetric and antisymmetric modes which are determined by the imaginary parts of the Green functions $-2\mathrm{Im}G^{(\pm)}_{AB}(\omega)$, where the Green functions $G^{(\pm)}_{AB}(\omega)$ are defined as $G^{(\pm)}_{AB}(\omega)=\langle\langle\tilde{b}_{A\nu},\tilde{b}_{B\nu};\tilde{b}^{+}_{A\nu}\pm\tilde{b}^{+}_{B\nu}\rangle\rangle$. The plus (+) and (-) signs refer to the symmetric and antisymmetric combination with respect to the interchange between the atom A and its image B; $G^{(\pm)}_{AB}(\omega)$ are the Fourier transforms of the corresponding time dependent Green functions. The expressions for the absorption coefficient describing the transition $|0\rangle\leftrightarrow|\nu\rangle$ for the symmetric and antisymmetric modes of the system consisting of the atom A interacting with its image B are determined by the spectral functions which are obtained from the imaginary parts of the Green functions $G^{(\cdot\cdot)}_{AB}(\omega)$, namely $-2\mathrm{Im}G^{(\pm)}_{AB}(\omega)$.

3. Excitation Spectra

To discuss the excitation spectrum, we consider the case of short distances of separation between the atom and the metal surface, i.e., when the equality $(2RE_{\nu 0}/c)<1$ is satisfied and retardation effects are unimportant [10-14,16, 17]. Using the Hamiltonian (1) we derive the following equations of motion for the Green functions [10,11]

$$G_{AB}^{(\pm)}(\omega) = \frac{(\bar{n}_0 - \bar{n}_\nu)}{\pi} 2E_{\nu0}\left[\omega^2 - \omega_{sp}^2(1\;\alpha) + i\omega\gamma_{sp}\right]\{(\omega^2 - \omega_{j+}^2)(\omega^2 - \omega_{j-}^2) + i\omega\gamma_{sp}(\omega^2 - E_{\nu0}^2)$$
$$+ \frac{i\gamma_0}{2}\left[1\pm\varepsilon_1(\omega)\right]^2\left[\omega^2 - \omega_{sp}^2(1\;\alpha) + i\omega\gamma_{sp}\right]E_{\nu0}\}^{-1} , \tag{2}$$

which describe the Green functions of the symmetric (+) and antisymmetric (-) modes of the system for j=s and j =a, respectively. The functions $\bar{n}_0 = \langle\alpha_{A0}^+\alpha_{A0}\rangle = \langle\alpha_{B0}^+\alpha_{B0}\rangle$ and $\bar{n}_\nu = \langle\alpha_{A\nu}^+\alpha_{A\nu}\rangle = \langle\alpha_{B\nu}^+\alpha_{B\nu}\rangle$ denote the average values of the electron number operators describing the electron population of the states $|0\rangle$ and $|\nu\rangle$ at t=t', respectively. The energies of excitation $\omega_{j\pm}$ are determined by

$$\omega_{j\pm}^2 = \tfrac{1}{2}\left\{E_{\nu0}^2 + \omega_{sp}^2 \pm\left[(E_{\nu0}^2 - \omega_{sp}^2)^2 \pm 4E_{\nu0}^2\omega_{sp}^2\alpha\right]^{\frac{1}{2}}\right\} , \tag{3}$$

where $\omega = \omega_{s\pm}$ (j=s and the plus sign in the square bracket) and $\omega = \omega_{a\pm}$ (j=a and the minus sign in the square bracket) refer to the symmetric and antisymmetric modes, respectively. The coupling function α is defined [14] as

$$\alpha = \alpha(R) = \alpha_s/2R^3 , \tag{4}$$

where α_s is the static polarizability of the atom and ω_{sp} is the frequency of the surface plasmons defined as $\omega_{sp} = \omega_p/\sqrt{2}$. The expressions (2) describe the Green functions of the symmetric and antisymmetric modes, where the effects of the damping for the surface plasmons are fully included. If energy shifts proportional to $\gamma_0\gamma_{sp}$ are neglected, as being very small, the excitation spectra of the symmetric and antisymmetric modes are determined by

$$J_{AB}^{(\pm)}(\omega) = -2\mathrm{Im}G_{AB}^{(\pm)}(\omega) = \frac{(\bar{n}_0 - \bar{n}_\nu)}{\pi}\left[I_j(\omega) + I_j(-\omega)\right] , \tag{5}$$

where the spectral functions $I_j(\pm\omega)$ are defined as

$$I_j(\omega) = \frac{N_j(\omega_{j+})\Gamma_\pm(\omega_{j+}) - (\omega-\omega_{j+})M_j\gamma_{sp}}{(\omega-\omega_{j+})^2 + \Gamma_\pm^2(\omega_{j+})} + \frac{N_j(\omega_{j-})\Gamma_\pm(\omega_{j-}) + (\omega-\omega_{j-})M_j\gamma_{sp}}{(\omega-\omega_{j-})^2 + \Gamma_\pm^2(\omega_{j-})} , \tag{6}$$

$$N_j(\omega_{j\pm}) = \pm\left(\frac{2E_{\nu0}}{\omega_{j\pm}}\right)\frac{\left[\omega_{j\pm}^2 - \omega_{sp}^2(1\mp\alpha)\right]}{(\omega_{j+}^2 - \omega_{j-}^2)} , \quad M_j = \frac{2E_{\nu0}}{(\omega_{j+}^2 - \omega_{j-}^2)} , \tag{7}$$

$$\Gamma_\pm((\omega_{j\pm}) = \gamma_\pm(\omega_{j\pm}) + \Gamma_{sp}(\omega_{j\pm}) , \tag{8}$$

$$\gamma_\pm(\omega_{j\pm}) = \frac{\gamma_0}{8} N_j(\omega_{j\pm})\left[1\pm\varepsilon_1(\omega_{j\pm})\right]^2 , \tag{9}$$

$$\lambda_{sp}(\omega_{j\pm}) = \Gamma_{sp}(\omega_{j\pm})/\gamma_{sp} = \pm\tfrac{1}{2}\frac{(\omega_{j\pm}^2 - E_{\nu0}^2)}{(\omega_{j+}^2 - \omega_{j-}^2)} . \tag{10}$$

The spectral function $I_j(\omega)$ describes asymmetric Lorentzian lines, which are peaked at the frequencies $\omega = \omega_{j\pm}$ having spectral widths of the order of $\Gamma_\pm(\omega_{j\pm})$; the extent of the asymmetry at $\omega\neq\omega_{j\pm}$ depends on the strength of the damping of the surface plasmons γ_{sp}. The first term in the expression (8) describes the radiative damping, and the second term represents the damping of the surface plasmons, while terms proportional to $\gamma_0\gamma_{sp}$ have been neglected. However, in the presence of plasmon damping, since γ_{sp} is always much greater than γ_0, $\gamma_{sp}\gg\gamma_0$ (or the corresponding lifetimes $\tau_0\gg\tau_{sp}$), the radiative one-atom-image process is erased by the nonradiative process. Hence, the radiative and nonradiative processes compete with each other only when the radiative processes arise from cooperative effects of an assembly of N identical atoms adsorbed on the metal surface. In this case, the effective radiative damping [10,11,16] is equal to $N\gamma_0$ for large N and may be in a position to compete with γ_{sp}.

4. Nonradiative Excitation Spectra

In the case when the surface plasmon mode is heavily damped, namely, when $\gamma_{sp}\gg\gamma_0$, which is always valid, the radiative contributions to the damping can be neglected, and $\Gamma_\pm(\omega_{j\pm})\approx\Gamma_{sp}(\omega_{j\pm})$ is defined by (10). In this case nonradiative energy transfer of the excitation energy of the atom to the metal takes place, and in this limit the spectral functions (6) are reduced to [10]

$$\gamma_{sp}I_j(\omega) = \frac{N_j(\omega_{j+})\lambda_{sp}(\omega_{j+})-X_{j+}M_j\gamma_{sp}}{[X_{j+}]^2 + \lambda^2_{sp}(\omega_{j+})} + \frac{N_j(\omega_{j-})\lambda_{sp}(\omega_{j-})+X_{j-}M_j\gamma_{sp}}{[X_{j-}]^2 + \lambda^2_{sp}(\omega_{j-})} , \qquad (11)$$

where $X_{j\pm}=(\omega-\omega_{j\pm})/\gamma_{sp}$, defines the reduced frequency. The spectral function $\gamma_{sp}I_j(\omega)$ in units of γ_{sp} describes asymmetric Lorentzian lines, which are peaked at the frequencies $\omega=\omega_{j\pm}$ and having spectral widths of the order of $\gamma_{sp}\lambda_{sp}(\omega_{j\pm})$. When $E_{\nu 0}$ is greater than ω_{sp}, $E_{\nu 0}>\omega_{sp}$, the excitations at $\omega=\omega_{s+}(\omega_{a+})$ and at $\omega=\omega_{s-}(\omega_{a-})$ are referred to as the atomic-like and surface plasmon-like excitations of the symmetric (antisymmetric) modes, respectively; the reverse is true for $\omega_{sp}>E_{\nu 0}$.

In our recent study [10,11) use has been made of (11) to prove through a numerical calculation that the optical spectra of Xe on Al and those of Kr on Au and on Mg vanish while for others the spectra persist. We shall make use of (11) to study numerically the optical spectra of Xe on the surfaces of Ti and Au as well as those of Ar on Mg and on Au for which experimental data are available [1-3]. Predictions have also been made in [18] for the adsorption spectra of excited Xe and Ar on the metal surfaces of W, Ni and Rh, respectively, at various distances R from the atom to the corresponding metal surfaces.

5. Numerical Results

For given values of R and for the values of $\alpha_s=4A^3$ and $1.6A^3$ for Xe and Ar [14], respectively, the corresponding values of α, can be determined from (4). Using (6)-(10) and the computed values of α, we derive the numerical data for the spectra of (11) for two excited states of Xe [19] physisorbed on Ti and Au, respectively; and for two frequencies of Ar on Mg and Au. The observed values [20,21] for the surface plasmon frequencies $\omega_{sp}=15.556$, 7.4953 and 17.1827 eV have been used for Ti, Mg and Au, respectively. Using these data, the computed spectra from (11) for the atomic-like excitations are illustrated in Figs. 1-4, where the relative intensity per atom $I_j(\omega)$ in

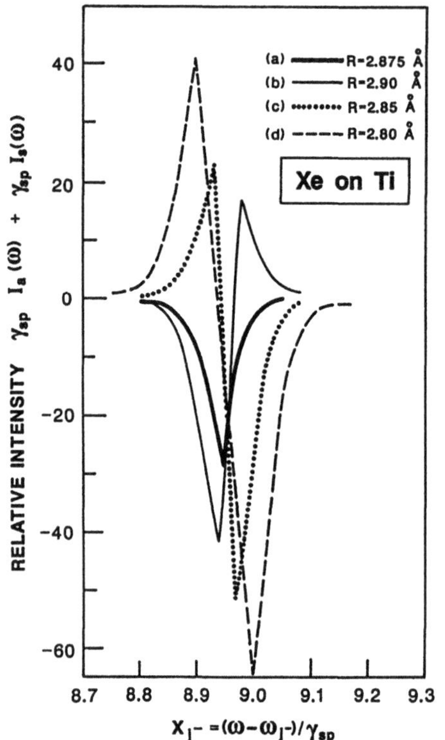

Fig. 1. Spectra of the atomic-like excitations of Xe atoms at the frequencies $E_{v0} = 8.437$ eV and 9.57 eV, which are physisorbed on the surface of Ti at various distances R from the surface of the metal.

Fig. 2. As in Fig. 1 but for excited Xe on the surface of Au.

units of γ_{sp}, $\gamma_{sp}I_j(\omega)$, is plotted versus the relative frequency $X_{j\pm} = (\omega-\omega_{j\pm})/\gamma_{sp}$ in units of eV/γ_{sp} for the symmetric j=s and antisymmetric modes j=a, respectively. Results of numerical calculations of the spectral functions (11) indicate that the intensities of the symmetric and antisymmetric modes take positive and negative values, respectively. The graphs in Figs. 1-4 illustrate the net lineshapes, which result from the combination of the red(blue) shifted peaks of the symmetric (positive intenstities) modes and the blue(red) shifted peaks of the antisymmetric (negative intensities) modes of the two atomic excitations under investigation. Thus, each line in Figs. 1-4 represents the net result of the cancellation process, which corresponds to an excited geometrical configuration for a given value of R for the physisorbed atom in question.

The spectra of the atomic-like excitations of Xe on Ti and on Au are illustrated in Figs. 1 and 2, where the net lineshapes arising from the cancellation processes for the four excited configurations corresponding to different distances R are denoted by (a)-(d), respectively. The lineshapes

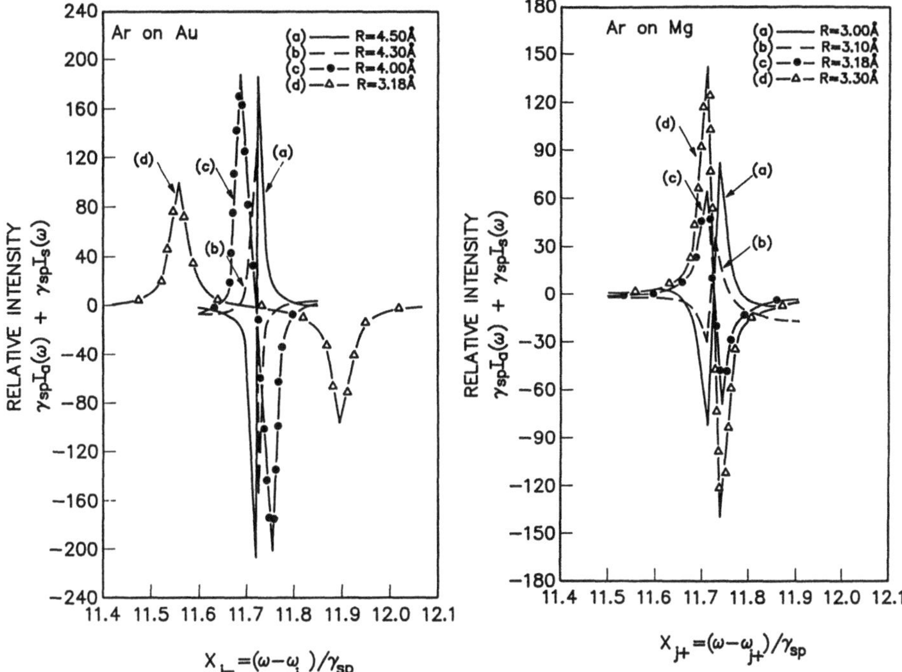

Fig. 3. As in Fig. 1 but for excited Ar at the frequencies $E_{\nu 0}$=11.62 eV and 11.83 eV, which one physisorbed on the surface of Au, respectively.

Fig. 4. As in Fig. 3 but for excited Ar on the surface of Mg.

in Figs. 1 and 2 describe absorption-amplification and amplification- absorption type of spectra, which absorb and amplify and vice-versa at the corresponding frequencies. When the absorption and the amplification frequencies are nearby, an infinitesimal perturbation or a very small disturbance in the proper direction may cause the disappearance of the unstable configuration and, consequently, the vanishing of the shape function in question. Inspection of Fig. 1 reveals that the (c) and (d) configurations describe absorption-amplication type spectra while the spectrum of the (b) configuration is of the opposite type, namely, the amplification-absorption one. Since the spectrum changes from the positive-negative type to that of the negative-positive one, the vanishing configuration should occur somewhere between those of the (b) and (c) configurations, namely, for values of R between R=2.90A and R=2.85A. The (a) configuration consists of a peak having a very small negative intensity, which implies amplification and being located between the (b) and (c) configurations as having the appropriate value of R equals to R=2.85A. Hence, it appears that the (a) configuration is the most likely one to be close to the vanishing one. Similar reasoning is applicable to Fig. 2 for the spectra of Xe on Au, where the (a) configuration with R=2.80A is the most probable one to be near the vanishing configuration. The observed spectra of Xe on Ti and on Au vanish [1-3]; therefore, our results are compatible with the experimental ones provided that Xe is physisorbed on Ti and on Au at distances R near R=2.875A and 2.80A, respectively.

Figures 3 and 4 illustrate the net lineshapes of Ar on the surface of Au and Mg, respectively, as a function of the distance R from the atom to the

corresponding metal surface. The (a) configuration in Fig. 3 illustrates
spectral lines, which are the opposite type from those described by the (b)-
(d) configurations and, hence if there is a vanishing configuration, it
should be between the (a) and (b) configurations with a suitable R taking the
values between R=4.5A and R=4.3A. Similarly, Fig. 4 indicates that the (a)
and (b) configurations describe spectra, which are opposite of the type than
those of the (c) and (d) ones; therefore, the vanishing configuration, if
there is any, should be between those of (b) and (c) with the appropriate
value of R between those of 3.10A and 3.18A. However, the spectral lines at
low coverage of Ar on Au and on Mg show strong persistence [1-3], which
implies that Ar is physisorbed on Au and on Mg at distances R other than
those indicated by the computed vanishing configurations.

6. Concluding Remarks

The lineshapes of the spectral lines of neutral excited atoms, which are
physisorbed on metal surfaces have been calculated. It is shown that at low
coverage and when the damping of the surface plasmons dominates the
occasional disappearance of the peaks of the spectral lines is due to the
existence of the symmetric and antisymmetric states, whose relative intensities take positive and negative values, respectively. Hence, they may have
a chance to cancel each other out whenever the frequency profiles of the
peaks in question coincide. The disappearance of the optical spectra of Xe
on Ti and on Au as well as the persistence of the corresponding spectra of Ar
on Au and on Mg is compatible with the observed ones [1-3]. It is suggested
that the disappearance or the persistence of the excited optical configurations may help to determine the distance R at which the atom is physisorbed
on the surface of the metal.

7. References

1. J.E. Cunningham, D.K. Greenlaw, J.L. Erskine, R.P. Layton, and C.P.
    Flynn, J. Phys. F: Metal Phys. 7, L281 (1977); J.E. Cunningham, D.K.
    Greenlaw and C.P. Flynn, Phys. Rev. Lett. 42, 328 (1979); J.E.
    Cunningham, D.K. Greenlaw, and C.P. Flynn, Phys. Rev. B22, 717 (1980);
    J.E. Cunningham, D. Gibbs, T.H. Chiu, and C.P. Flynn, J. Phys. C: Solid
    State 14, L1113 (1981).
2. C.P. Flynn, and Y.C. Chen, Phys. Rev. Lett. 46, 447 (1981); C.P. Flynn,
    and J.E. Cunningham, J. Phys. C: Solid State 15, L1169 (1982).
3. D. Gibbs, J.E. Cunningham, and C.P. Flynn, Phys. Rev. B29, 5292 (1984);
    J.E. Cunningham, D. Gibbs, and C.P. Flynn, Phys. Rev. B29, 5304 (1984);
    Y.C. Chen, J.E. Cunningham, and C.P. Flynn, Phys. Rev. B30, 7317 (1984);
    C.P. Flynn, Surface Sci. 158, 84 (1985).
4. N.D. Lang, A.R. Williams, F.J. Himpsel, B. Reihl, and D.E. Eastman,
    Phys. Rev. B26, 1728 (1982); T.-C. Chiang, C. Kaindl and D.E. Eastman,
    Solid State Commun. 41, 661 (1982).
5. J.E. Demuth, Ph. Avouris, and D. Schmeisser, Phys. Rev. Lett. 50, 600
    (1983); W. Eberhardt, and A. Zangwill, Phys. Rev. B27, 5960 (1983).
6. A. Bagchi, R.G. Barrera, and B.B. Dasgupta, Phys. Rev. Lett. 44, 1475
    (1980); A. Bagchi, R.G. Barrera, and R. Fuchs, Phys. Rev. B25, 7086
    (1982).
7. M. Tsukada and W. Brenig, Surface Sci. 151, 503 (1985).
8. Ph. Avouris and J.E. Demuth, Surface Sci. 158, 21 (1985).
9. H. Ueba and A. Yoshimori, Surface Sci. 175, 659 (1986).
10. C. Mavroyannis, Mol. Phys. 64, 457 (1988).
11. C. Mavroyannis, Can. J. Chem. 66, 741 (1988).

12. C. Mavroyannis, Mol. Phys. $\underline{6}$, 593 (1963).
13. C. Mavroyannis, and D.A. Hutchinson, Solid State Commun. $\underline{23}$, 463 (1977).
14. C. Mavroyannis, Mol. Phys. $\underline{36}$, 1565 (1978); Chem. Phys. Lett. $\underline{56}$, 263 (1978).
15. J.G. Endriz and W. Spicer, Phys. Rev. B$\underline{4}$, 4144 (1971).
16. R.H. Dicke, Phys. Rev. $\underline{93}$, 99 (1954).
17. C. Mavroyannis, Solid State Comm. $\underline{23}$, 467 (1977).
18. C. Mavroyannis, Applied Phys. A (in press).
19. C.E. Moore, <u>Atomic Energy Levels</u>, Natn. Bur. Stand. (U.S. Circ. No. 35, U.S. GPO, Washington, D.C., 1971).
20. G.G. Kleiman, and U. Landman, Phys. Rev. B$\underline{8}$, 5484 (1973).
21. B.E. Nieuwenhuys, O.G. Van Aardene and W.M.H. Sachtler, Chem. Phys. $\underline{5}$, 418 (1974).

The Interaction of Aromatic Molecules with the Basal Plane of Graphite and Rare Gas Atoms

G. Vidali[1] and M. Karimi[2]

[1]Syracuse University, Physics Department, Syracuse, NY 13244, USA
[2]Utica College, Physics Department, Utica, NY 13502, USA

1. Abstract

We have calculated the interaction of large aromatic molecules with the basal plane of grapite. We summed a Lennard-Jones pair potential with parameters determined by atom beam scattering and calorimetry data. We report here on the binding energies and vibrational frequencies of these molecules on graphite.

We have also calculated the interaction of rare-gas atoms with benzene and coronene. Specifically, we present a new model for the rare-gas/aromatic molecule interaction which gives a He-benzene equilibrium distance in agreement with molecular beam experiments; the He-C potential so determined, when applied to He-graphite, fits atom beam scattering data very well.

2. Introduction

The study of the interaction of large aromatic molecules with surfaces is interesting for two reasons. First, it gives the opportunity to study a variety of issues in the chemistry at solid surfaces, such as the weakening of intramolecular bonds due to adsorption, associative/dissociative adsorption/desorption processes and the influence of substrate geometry on adsorption. Second, recently it has been proposed that polycyclic aromatic hydrocarbons are present in the interstellar medium [1]. It is then of interest to study the interaction of these molecules with surfaces simulating dust grains.

There have been a few studies of the interaction of benzene with metal surfaces using surface science probes (see Ref. [2] and references cited therein). There are some common features in the interaction of benzene with smooth metal surfaces: 1) the adsorption doesn't greatly affect intramolecular vibrations, and 2) benzene lies flat on the surface.

While there has been much experimental activity in this area due to the increasing availability of a host of surface science probes [2], which can be used _in-situ_, theoretical understanding of the interaction and adsorption/desorption processes have not advanced at the same pace. Here we want to report on preliminary work that we have done recently on the interaction of large aromatic molecules with surfaces. In particular, we want to explore the possibility of extending theoretical models and methods that have been successfully applied to the interaction of atoms or simple molecules with surfaces [3]. In doing so we also expect to gain new insights in how to treat this more challenging problem.

Springer Series in Surface Sciences, Vol. 17
Adhesion and Friction   Editors: M. Grunze and H.J. Kreuzer
© Springer-Verlag Berlin, Heidelberg 1989

We have chosen to study the interaction of benzene and coronene with the basal plane of graphite. There are two main advantages for this choice: first, the basal plane of graphite is rather unreactive and thus many complications associated with chemisorption bonds can be avoided; second, the carbon-carbon and carbon-hydrogen interactions have been studied for many years and are better known than for many other systems [3,4].

Another interesting related problem that we wanted to explore was the reliability of the effective He-carbon potential [5] which fitted so well He-graphite [6] and He-diamond data [7,8]. Therefore, we decided to use the He-carbon potential so obtained to study the interaction of He with benzene and coronene. In particular, we wanted to see whether, as reported previously [9], this potential was inadequate to predict the He-benzene equilibrium distance [10]. Subsequently, the measured energy levels of He-graphite were revised upwards [11] and a new theoretical framework, essentially based on Effective Medium Theory (see below), to treat the He-graphite interaction was developed [12].

3. The Model

a) Rare-gas atoms/benzene or coronene

We have computed the interaction between rare gas atoms and benzene or coronene using two different methods. In the first (to be referred as Model 1), the interaction potential is written as a sum of Lennard-Jones potentials. The parameters σ=2.74 Å and ϵ=1.4 meV for He-C were taken from Ref. [5]. In that paper it was shown that by using a Lennard-Jones potential for He-C, one could build a He-graphite interaction potential that fitted remarkably well atom beam scattering and thermodynamic data [6]. These He-C parameters are somewhat different from the ones obtained using standard combination rules. For a discussion of this point, please see Refs. [3,5]. For the He-H we used parameters reported in Ref. [4]: σ=2.99 Å, ϵ=0.64 meV. Parameters for other rare gases interacting with C and H are reported in Table 1.

In the second method (to be referred as Model 2), the attractive part of the interaction is described by damped induced dipole and quadrupole terms. The repulsive part uses the so-called Effective Medium Theory which relates the surface charge density to the repulsive part of the potential via a proportionality constant α_0 [13]. We have applied this formalism successfully to many systems [14,15]. A complete description of

Table 1. Lennard-Jones parameters from Ref. [22,23].

System	σ(Å)	ϵ (meV)
Ne-C	2.80	2.65
Ar-C	3.10	6.17
Kr-C	3.21	7.38
Xe-C	3.36	8.93
Ne-H	2.78	2.82
Ar-H	3.15	4.73
Kr-H	3.27	5.99
Xe-H	3.40	6.65

this method has been published before [14]; here we will give a brief description.

The repulsive part of the potential is written:

$$V_R(\vec{r}) = \alpha_0 \rho(\vec{r}),$$

where ρ (r) is the charge density of the graphite surface [12,16]. For computational convenience we write:

$$\rho(\vec{r}) = \Sigma_i \; \rho(\vec{r}_i)$$

where $\rho_i(\vec{r}_i) = (A/r_i) \exp(-\gamma r_i)$ and where r_i is the position of the carbon atom; $A=10.83$ $\mathring{A}^{-2}$ and $\gamma=3.67$ $\mathring{A}^{-1}$.

The attractive part is given by a sum of damped dispersion terms between the adatom and the atoms in the solid. The induced dipole interaction contains terms proportional to C_6/r_6, while the induced quadrupole interaction is approximated by an expression proportional to C_8/r_8. The complete expressions are given in Refs. [14,15].

b) Benzene and coronene on graphite

As it will be discussed in the next section, the second model which uses EMT is to be preferred to the summation of Lennard-Jones potentials. However, its application is restricted to rare-gas atoms and a few others for which explicit calculations of α_0 exist. In the case of C-C or C-H interactions, these coefficients α_0 are not available [17]. We are therefore forced to use summation of Lennard-Jones potentials. On the positive side, the potential parameters are quite well known since many calculations have been done for graphite [3]. The parameters were obtained from Refs. [4,18]; for C-C we have $\sigma=3.4$ $\mathring{A}$, $\epsilon=2.41$ meV, while for C-H $\sigma=2.98$ $\mathring{A}$ and $\epsilon=1.46$ meV.

For the interaction distances of interest here, the substrate looks quite smooth. Obviously, the substrate will have an influence in locking the adsorbed layers in certain positions. In fact, commensurate phases are observed for certain coverages of benzene on metal surfaces and graphite [2]. We have calculated the lateral average of the C-graphite and H-graphite interactions and then summed all the contributions from the C and H atoms:

$$V_{00}(z) = \Sigma_i \left[V_i^C(z) + V_i^H(z) \right]$$

In the case of coronene we have omitted the H-graphite interaction since it contributes already by less than 2% to the well depth in the case of benzene/graphite.

The potential so obtained is inserted in the Schrodinger equation and the wave functions, position expectation values and bound states of the molecule are computed. To obtain the vibrational frequencies of benzene and coronene on graphite we approximated the bottom of the potential well with a harmonic oscillator well:

$$V_{00}(z) \simeq -D + \frac{1}{2} k \; (Z-Z_{min})^2$$

from which the vibrational frequency can be obtained.

$$(\nu = \frac{1}{2\pi} \sqrt{\frac{k}{m}})$$

4. Results and Discussion

We first tested our He-carbon potential built using EMT (Model 2) by computing the He-graphite potential. We obtained the following bound states (in parenthesis are the experimental values from Ref. [11]): 12.33 (12.42), 6.65 (6.59), 3.17 (3.05), 1.30 (1.24), 0.11 (0.16) meV. To obtain these results we had to decrease C_6 coefficients of the dispersion interaction by 5%. This is not surprising since C_6 values are usually not very well known.

The expectation value of the position of He above the basal plane is: 3.15 Å. Previous determinations using a potential based on EMT (somewhat similar to ours) gave [12]: 3.1 Å, while a summation of Lennard-Jones potentials gives: 2.9 Å. A determination of <z> based on neutron scattering data yields [19]: 2.85 Å. We tend to believe more in the EMT based potentials than in the Lennard-Jones since the former construction is based on physical parameters rather than on analytically convenient terms. As the expectation value of z is concerned, Lennard-Jones based potentials show an increase of <z> going from He to Xe, while viceversa happens using EMT potentials, except for Ne [12, 14, 15].

The results of our calculation for He, Ne and Ar interacting with benzene and coronene and using the two models introduced in the previous section are reported in Table 2. We would like to point at two results. First, as expected, the deepest binding site is the one above the center of the benzene ring or in the center of coronene. The binding energy for rare gases on aromatic molecules increases as the size of the molecule gets bigger and approaches the basal plane of graphite. Second, the expectation value of the position z for He/benzene, 3.1 Å, (using Model 2) is in good agreement with the value of 3.17 ± 0.37 Å determined experimentally [10]. Overall we assign a higher degree of confidence on Model 2; however, the difference in binding energies between the two potentials is usually small, except for Ar.

We next comment on the results for benzene/graphite and coronene/graphite (see Table 2). The binding energy per carbon atom is about 45 meV, comparable to an experimental determination for separating two graphite planes (23 meV) [20] or to recent density functional calculations for graphite (100 meV) [21]. The distance between benzene and graphite is 3.38 Å, i.e. of the order of the interplanar distance in graphite. The vibrational energies for benzene and coronene with respect to the basal plane are about 45 meV in both cases. In the case of benzene, this value can be compared to the one for benzene adsorbed on a metal surface. In the case of Ru(001), a recent experiment (2) gave 36 meV, in agreement with the fact that, the interaction of an atom or molecule with the basal plane of graphite is stiffer than with a metal surface [22].

Table 2.

Key parameters of rare-gas/benzene or coronene and benzene, coronene/graphite interactions. D, <z>, E_0 and E_i are the well depth,expectation value of adatom position and first two bound states.V_A, V_B and V_C are the minima of the potential at different adsorption sites: A,center of hexagon; B, on top of a C atom; and C,between two C atoms. Energies are in meV and lengths are in Å.

System	D	<z>	E_0	E_1	V_A	V_B	V_C
He-benz[a]	11.5	2.9	7.7	2.7	11.5	7.8	8.5
	12.0	3.1	8.3	3.1	12.0	8.7	9.3
Ne-benz[a]	25.9	2.8	23.0	17.7	25.9	19.2	20.4
	29.5	2.8	26.5	21.0	29.5	22.1	23.5
Ar-benz[a]	57.5	3.1	54.3	48.2	57.5	44.2	46.7
	104.8	2.7					
Kr-benz	71.4	3.3	68.6	63.1	71.4	55.7	58.8
Xe-benz	84.3	3.4	81.6	76.3	84.3	67.6	70.9
He-Cor	15.8	2.8	11.5	5.4	-	-	-
Benz-gr	277.6	3.38	274.1	260.4	-	-	-
Cor-gr	1103.0	-	-	-	-	-	-

[a] The top line is obtained with Model 1 (Lennard-Jones potential); bottom line is with Model 2 (EMT).

5. Conclusions

The study of the interaction of large aromatic molecules with surfaces is just at the beginning. The theoretical modelling of such interaction is particularly challenging because of the number of degrees of freedom involved. While it would be desirable to do ab-initio calculations, we have quite a long way to go to achieve that goal. On the other hand, some methods that have been successfully employed in the study of atoms or simple molecules interacting with surfaces can be used to study the interaction of larger molecules with surfaces. This is the path that we chose to follow.

We have shown that key parameters of the molecule-surface interaction can be obtained quite easily. Furthermore, as demonstrated in the case of He-benzene, one can test the accuracy of the He-C potential as determined from He-graphite and He-diamond data; this test suggests that EMF based potentials give an overall better description of the interaction. We are planning to extend this type of model to more complicated interactions.

<u>6. Acknowledgements</u>

One of us (G.V.) would like to acknowledge the Alfred P. Sloan Foundation for partial support of this work.

<u>7. References</u>

1. See articles in <u>Polycyclic Aromatic Hydrocarbons and Astrophysics</u>, Eds. A. Leger, L. d'Hendecourt, and N. Boccara (D. Riedel Publishing Co., Boston, 1987).
2. P. Jakob and D. Menzel, Surf. Sci. <u>201</u>, 503 (1988); C. Bondi and G. Taddie, Suf. Sci. <u>203</u>, 587 (1988). J. C. Bertolini and J. Massandier in the Chemical Physics of Solid Surfaces and Heterogeneous Catalysis, Ed. D. A. King and D. P. Woodruff (Elsevier, New York, 1984), V.3, Ch. 3, p. 107.
3. W. A. Steele, <u>The interaction of gases with solid surfaces</u> (Pergamon, Oxford, 1974); H. Hoinkes, Rev. Mod. Phys. <u>52,</u> 933 (1980).
4. J. M. Phillips and M. D. Hammerbacher, Phys. Rev. B<u>29</u>, 5860 (1984) and references cited therein.
5. W. E. Carlos and M. W. Cole, Phys. Rev. Lett. <u>43</u>, 697 (1979); Surf. Sci. <u>91</u>, 339 (1980).
6. Atom beam scattering and thermodynamic data are summarized in: M. W. Cole, D. R. Frankl, and D. L. Goodstein, Rev. Mod. Phys. <u>53</u>, 199 (1981).
7. G. Vidali and D. R. Frankl, Phys. Rev. B. <u>27</u>, 2480 (1983).
8. G. Vidali, M. W. Cole, W. H. Weinberg, and W. A. Steele, Phys. Rev. Lett. <u>51</u>, 118 (1983).
9. S. Leutwyler and J. Jortner, J. Phys. Chem. <u>91</u>, 5588 (1987).
10. S. M. Beck, M. G. Liverman, D. L. Monts, R. J. Smalley, J. Chem. Phys. <u>70</u>, 232 (1979).
11. S. Chung, A. Kara, and D. R. Frankl, Surf. Sci. <u>171</u>, 45 (1986); J. C. Ruiz, G. Scoles, and H. Jonsson, Chem. Phys. Lett. <u>129</u>, 139 (1986).
12. F. Toigo and M. W. Cole, Phys. Rev. B<u>32</u>, 6989 (1985).
13. J. K. Norskov, Phys. Rev. B<u>26</u>, 2875 (1982); N. D. Lang and J. K. Norskov, Phys. Rev. B<u>27</u>, 4612 (1983). M. W. Cole and F. Toigo, Phys. Rev. B<u>31</u>, 727 (1985).
14. M. Karimi and G. Vidali, Phys. Rev. B<u>38</u>, 7759 (1988); in <u>Diffusion at Interfaces: Microscopic Concepts,</u> Eds. Kreuzer, Weimer, and Grunze Springer Series in Surface Science, V. 12, 43 (1988).
15. M. Karimi and G. Vidali, Phys. Rev. B. 1988 (in press).
16. M. W. Cole, private communication.
17. See also Ref. 13 and papers cited therein. In general V_R is not linearly related to $\rho(r)$ and in fact in V_R is a complicated function of ρ.
18. W. A. Steele, J. Phys. Chem. <u>82</u>, 817 (1978).
19. K. Carneiro, L. Passell, W. Thomlinson, and H. Taub. Phys. Rev. B. <u>24</u>, 1170 (1981).
20. L. A. Girifalco and R. A. Lad, J. Chem. Phys. <u>25</u>, 693 (1956).
21. D. P. DiVincenzo, E. J. Mele, and N. A. W. Holzwarth.
22. G. Vidali, M. W. Cole, and J. R. Klein, Phys. Rev. B<u>28</u>, 3064 (1983).
23. J. P. Toennies, W. Welz, and G. Wolf, J. Chem. Phys. <u>71</u>, 614. (1979).

Index of Contributors

MIX
Papier aus verantwortungsvollen Quellen
Paper from responsible sources
FSC® C105338

If you have any concerns about our products,
you can contact us on
ProductSafety@springernature.com

In case Publisher is established outside the EU,
the EU authorized representative is:
Springer Nature Customer Service Center GmbH
Europaplatz 3, 69115 Heidelberg, Germany

Printed by Libri Plureos GmbH
in Hamburg, Germany